HANDCRAFTED CERAMIC TILES

HANDCRAFTED CERAMIC TILES

Mike Jones
and
Janis Fanning

Sterling Publishing Co., Inc.
New York

A QUARTO BOOK

Library of Congress Cataloging-in-Publication-Data is available upon request.

Published by Sterling Publishing Co., Inc.
387 Park Avenue South
New York, NY 10016-8810

This book was designed and produced by
Quarto Publishing plc
The Old Brewery
6 Blundell Street
London N7 9BH

Distributed in Canada by Sterling Publishing
c/o Canadian Manda Group, One Atlantic Avenue, Suite 105
Toronto, Ontario, Canada M6K 3E7

Writers Mike Jones (pages 50-125), Janis Fanning (pages 10-45), Jill Blake (pages 6-9, 128-138)
Project editor Ann Baggaley
Text editors Ann Baggaley, Anthony Atha, Maggi McCormick
Designer Sheila Volpe
Photographers Verity Welstead, Paul Forrester, Charles and Patricia Aithie
Illustrator Kate Simunek
Picture researcher Miriam Hyman
Senior art editors Penny Cobb, Catherine Shearman
Managing editor Sally MacEachern
Editorial director Pippa Rubinstein
Art director Moira Clinch

Typeset in Great Britain by Central Southern Typsetters, Eastbourne
Manufactured in Singapore by Pica Colour Separation Overseas Pte Ltd
Printed in Singapore by Star Standard Industries (Pte) Ltd

ISBN 0-8069-9676-5

PUBLISHER'S NOTE
Pottery can be dangerous. Always follow the instructions carefully, and exercise caution. Follow the safety procedures accompanying the techniques.
As far as the methods and techniques mentioned in this book are concerned, all statements, information, and advice given here are believed to be accurate. However, neither the author, copyright holder, nor the publisher can accept any legal liability.

CONTENTS

MAKING TILES 48

MAKING AND DECORATING TILES 67

DECORATING BISCUIT-FIRED TILES 99

DECORATING PRE-GLAZED TILES 107

GALLERY 120

WORKING WITH TILES 126

INDEX 142

ACKNOWLEDGMENTS 144

INTRODUCTION

ABOVE Detail from a tiled panel in the church of Our Lady of El Prado at Talavera in Spain. This town was a leading center for the production of ceramics in the 17th century.

For thousands of years ceramic tiles have been used as a practical and decorative surface in architecture and interior design. Tiles are hard-wearing, long-lasting, waterproof, and heatproof; qualities that make them a natural choice for floors, roofs, fireplaces, kitchens, and bathrooms. In medieval times, beautifully ornamented and colored tiles were used to make decorative patterns around the fountains and courtyards of the East, while in the Western world today, tiles are often found in public buildings, restaurants, stores, hospitals, and swimming pools, where they have proved to be both practical and hygienic.

LEFT The magnificent tiled dome of Shir-Dor Madrasah in Samarkand. The Islamic decorative ceramic tradition, which developed from the 9th century onward, was to spread its influence throughout the world.

For centuries tiles have been laid out in geometric patterns, varying colored and plain tiles, and often tiles have been decorated by artists, thereby adding another dimension, transforming a functional square into a unique object of beauty. Many of the finest examples of decorative tilework in the Middle East can still be seen, and the tiles there are as fresh and colorful now as they were when they were created centuries ago.

Today the meaning of the word tile is wide-ranging. Tiles can be made from any type of material, marble, stone, clay, linoleum, vinyl, or carpet, which comes in a modular form and can be used to cover interior or exterior surfaces.

The History of Tiles

Tiles have a long history. They originated in the Near and Middle East where they have been used in buildings for almost 4000 years. There, where clay is the natural building material, tiles cut from the clay and dried in the sun or baked in a kiln made an ideal surface for roofs, keeping buildings cool in the heat of the day, and providing good insulation after sundown.

Originally, tiles were brown or terracotta, the color of the earth itself, but as potters became more knowledgeable and ambitious, tiles were glazed and decorated and laid out more elaborately in concentric and decorative patterns. As early as 1449 B.C. the Egyptians used rather sandy tiles with a surface glaze stained a verdigris blue-green with copper and, from their knowledge of glassmaking, they developed more sophisticated glazes and patterns which they applied to their pottery and architectural ceramics.

Handmade terracotta tiles, with their rich natural colors, cast a warm glow across a hall floor. They are not only extremely attractive but practical and hard wearing.

Blue was the most common color and the first decorated tiles were made by the Egyptians scratching patterns on the clay before the tiles were glazed. These indentations then absorbed more of the glaze and stood out as darker blue patterns against a paler, turquoise background. Highly decorative wall tiles, which date from around 1180 B.C., have been found in the Temple of Medinet Habu at Tell al Yehudia in the Nile Delta, and these show captured slaves, animals and imaginary beasts, symbolic signs, and ornaments.

Between the 13th and 6th centuries B.C. the Assyrians and Babylonians made earthenware wall tiles and bricks, which were decorated in patterns in richly colored glazes and at Susa, the palace of the Achaemenid Kings of Persia had decorated tile friezes in colored relief showing motifs of lions and mythical winged bulls.

A 13th-century luster painted tile from Kashan in Persia. Luster decoration, with its glittering metallic sheen, was used in mosques and palaces throughout the Middle East.

After the conquest of the Middle East by Alexander the Great between the years 334–325 B.C., the practice of this decorative technique died out, and it remained in abeyance for several centuries. It was revived during the rule of the Sassanian kings (C.E 221–641). The first designs and patterns at this time echoed those of the Achaemenid period, but they were applied to panels of stucco rather than wall tiles, and the panels were molded and carved with figures and stylized representations of plants. This type of decoration remained popular with Islamic people, and stucco was frequently used in conjunction with tiles in formal schemes.

In C.E 750 the center of the Islamic world moved from Mecca, where it was founded, eastward to Baghdad. There the influence of Persian craftsmen became predominant, and inspired by the wares imported along the Silk Route from China, the great Islamic decorative ceramic tradition started, which spread throughout Europe and later all over the world.

The most important development in the 8th century was the introduction of luster decoration. Tiles were first ground-glazed in opaque white. They were then painted in shades of yellow, red, and brown, and the decoration was enhanced by gold luster, produced by using a compound of silver and gold in a medium of liquid ocher. When fired this luster produced a metallic sheen which glittered in the sunlight and was used to enhance the decoration of mosques, temples, and palaces. This technique spread to Egypt where it was used on pottery rather than tiles, and in the 13th century an

Dutch Delftware typically depicts birds, animals, plants, and small scenes. The corner decorations are generally of three basic types, the one shown here being the "Ox head" corner.

important tile-making center was established at Kashan, south of Tehran in Persia. There luster-painted and relief tiles were produced that were used throughout Persia and the Middle East.

Early Tile-making in Europe

In Europe, during this period, tiles were far more mundane and less colorful. Decorative encaustic tiles, a special inlaying technique imported from Egypt, were sometimes laid on the floors of medieval monasteries and churches, but generally such tiles as existed were the natural color of the clay from which they were made. It was not until the 18th century, when young aristocrats made the Grand Tour, that the rich and exotic decoration of the Middle East was discovered, and it was around this time that the East India Company started to import furniture, earthenware, and china from Asia and the Orient.

The style of decoration used on Chinese porcelain inspired the Dutch tile makers of the period to produce the majolica tiles which are now best known as Dutch Delftware. The designs were often very simple, showing figures and handpainted motifs, and the tiles were used as panels, in churches, and to decorate fireplaces and windowsills in the home. To start with, majolica tiles were decorated in rich colors, but eventually the characteristic blue and white designs took over and this is how most people think of Delftware today. Tiles produced in Holland became so popular that the industry spread throughout Europe and then across the Atlantic to America.

A tiled panel in the Victorian style. Five-tile motifs such as this were commonly used as borders to fireplaces, once the focal point of the home.

The Development of Decorative Tiles in the 18th and 19th Centuries

Until the middle of the 18th century, the decoration of tiles was done by hand. This was a laborious and costly process, but in 1756 a printing technique was developed in Liverpool by John Sadler and Guy Green, which allowed the consistent reproduction of very fine detail. They transferred an image from a printing plate to a tile using a transfer tissue. These early images were usually reproduced in one color, often black, red or brown, and the designs were fired on a white glazed tile at relatively low temperatures. The first designs were simple landscapes or figures, for the early printing plates were blocks made of wood – similar to those used at the time for fabric printing. But, as printing developed, plates were made from copper, and the design was engraved or etched on the surface of the metal. This enabled complex, finer detail to be

Handcolored tiles in the Kate Greenaway style, by T & R Boote, who were among celebrated designers of the 19th century. These are from a set depicting the Four Seasons, with Spring shown above and Winter below.

reproduced, and as the manufacturing process became more sophisticated, so more colors were used in the designs.

The Industrial Revolution of the 19th century brought a vast expansion of building to house the newly wealthy middle classes throughout the world, as well as factories, meeting halls, and churches. The Victorian style was highly ornamental and ceramic tiles were often used to decorate the floors and walls of the new houses. Tiles were most often positioned in the entrance hall to make a grand statement, and from there they spread to verandas and terraces and into sunrooms, which were being added to house the exotic plants imported from tropical countries. Tiles used outside had to be frostproof, and as the practicality and durability of tiled floors became more apparent, the use of tiles spread throughout the house, generally buff and black tiles laid in a pattern, many of which can still be seen today.

As plumbing improved tiles started to be used in bathrooms where they provided a practical wall and floor surface, and it also became the fashion to decorate the fireplace with tiles. These were often highly decorative.

Irregularly shaped tiles make a striking checkerboard facade on a house in Vienna.

Decorative Tiles in the 20th Century

The Victorian Gothic style was copied throughout the world and as a result many churches, civic buildings, and large houses have geometrically tiled floors, as well as panels designed to suggest frescoes or wall hangings. As the century moved to a close the style of decoration changed from Victorian Gothic to Art Nouveau. This was followed by the geometric designs of Art Deco. Tiles followed fashion and those with Egyptian motifs proved particularly popular after Tutankhamun's tomb was discovered in 1922.

Between the world wars the interiors of many hotels and restaurants were decorated with tiles and "Cunard Style," inspired by ocean-going liners, became all the rage. At home people stopped using tiles on their hall floors, and the use of decorative tiles was confined mainly to fireplace surrounds, while plainer tiles were used in bathrooms and kitchens.

Tiles Today

People searching for good ceramic tiles today will find a wide choice of patterns and designs available, and many modern potters have designed original tiles. Many tiles are hand-decorated which give individuality to a work surface, wall or panel and tiles can also be designed to order. Tiles are also produced color coordinated with fabrics used for curtains and furniture. New materials and methods of production are constantly being investigated, and often these look to the past for their inspiration. For example the modern oilslick finishes reproduce the effect of Victorian lusterware, which in itself is an imitation of the lusterware tiles produced in 13th-century Persia.

However, if you are looking for true originality, you can design your own tiles and produce a tiled surface that is unique and individual to you. You just have to follow the steps and techniques illustrated in this book.

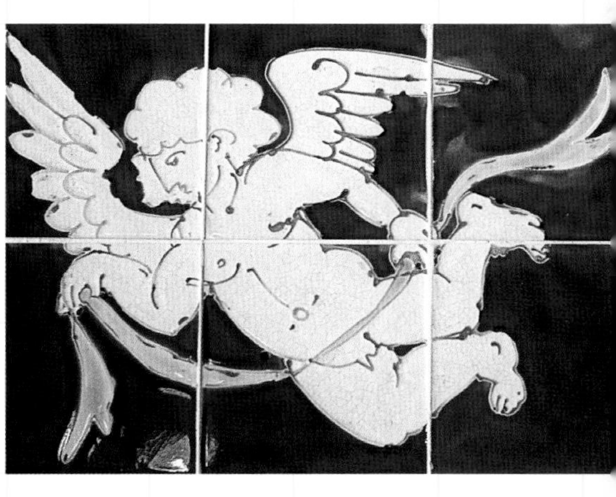

A decorative panel of handmade tiles created by a modern potter. Designers today can use the past for inspiration but with the advantage of today's new materials.

Tile Decoration

These projects use special paint effects, découpage, stencils, and transfer designs to decorate plain tiles. All the techniques are easy to follow and allow plenty of scope for creative expression.

Using a variety of simple craft techniques, you can transform a set of plain white pre-glazed tiles into a striking decorative feature. No firing is necessary for the following projects.

Tile Decoration

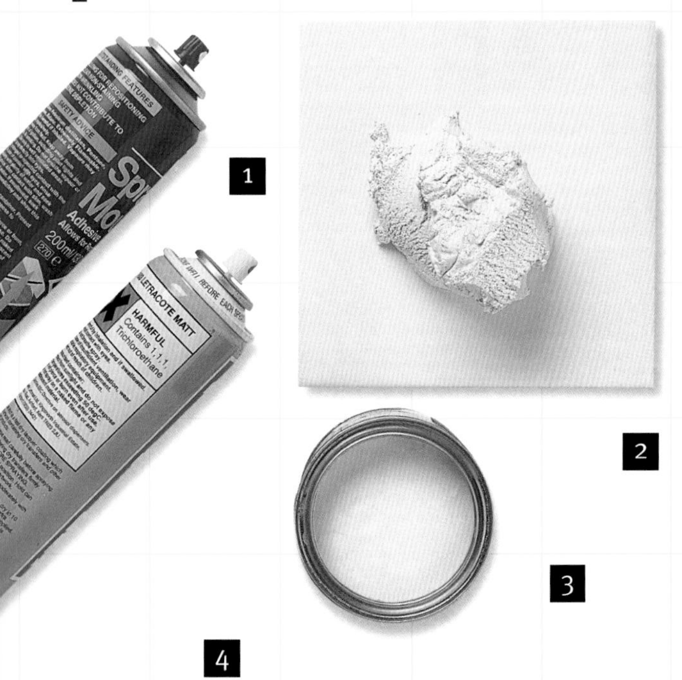

Tiles have adorned our surroundings, both interior and exterior, for centuries all over the world. Despite a colorful history, today's trends tend to lean toward a predominance of white. Although still very appealing in itself, a plain surface of tiles makes a perfect "canvas" on which the artisan can set to work.

Tile decoration is a most rewarding art with profound effect. Tiles can be transformed using simple techniques such as découpage, stenciling, painting, and transfer. Each craft creates a very different character, and some designs are more suited to a particular technique than others. However, no matter which technique you

1 Spray-on glue and lacquer; **2** grout; **3** white acrylic paint; **4** cutting mat; **5** tracing paper; **6** transfer paper; **7** permanent ink pen; **8** craft knife; **9** pencil; **10** mineral spirits; **11** colored acrylic paints.

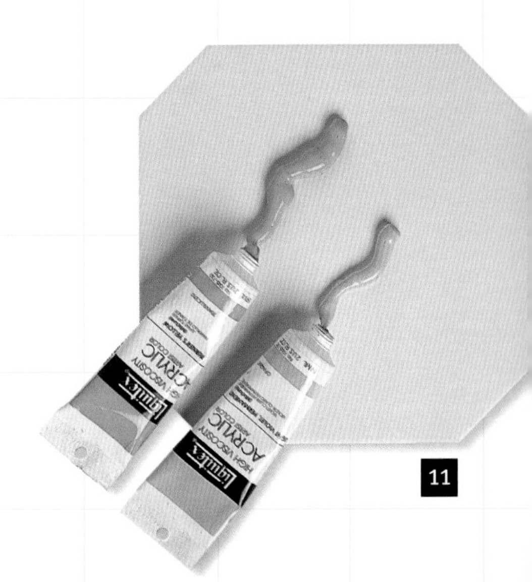

decide to use, the time and care taken during the project preparation will be reflected in the finished piece.

Consider the tile color, texture, and size when choosing the technique and design. Enhance what is already there. For instance, découpage works most effectively on a flat surface, either applied to an individual tile or divided between several. Stenciling is useful for repetitive designs, for example applying a border pattern on either textured or smooth finished tiles. Special paint effects are versatile, but because a glazed tile is not absorbent, a hard-wearing paint like enamel or acrylic must be used. Transfer paper is invaluable for transferring a design, and it can be used on either a rough or smooth surface tile – although the latter is preferable. The graphite backing is similar to a carbon copy in that it allows the most delicate of details to be reproduced again and again. Available from a good art supply store, transfer paper is a useful substitute if you lack confidence in drawing freehand.

When thinking about your design, look in magazines and books for inspiration. Research art movements, from Art Nouveau to Art Deco, looking at period styles from medieval to Victorian. Other sources for material are children's comics, old posters, recipe books, maps,

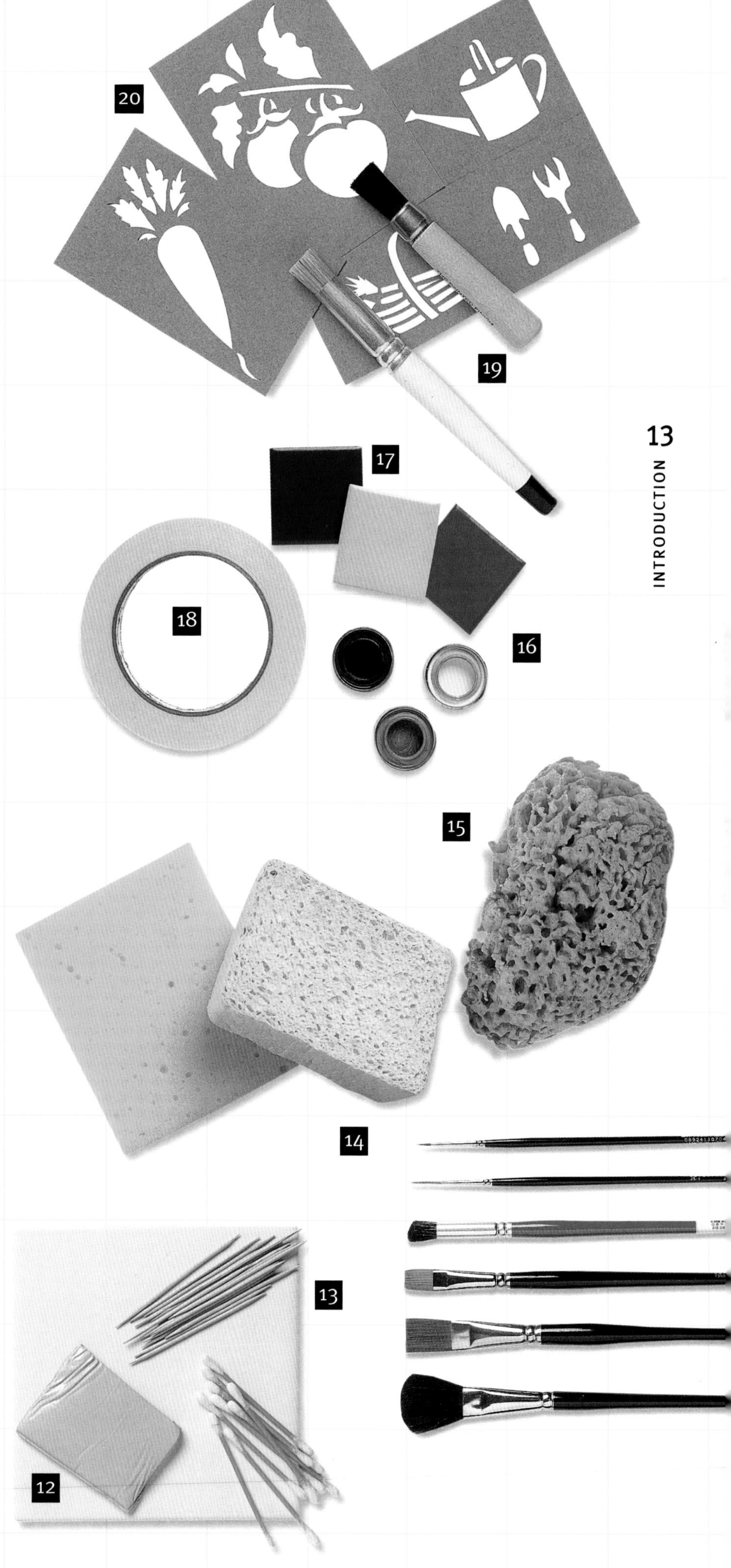

12 Removable adhesive; **13** cotton swabs and toothpicks; **14** paintbrushes; **15** synthetic and natural sponges; **16** enamel paints; **17** glazed tiles; **18** masking tape; **19** stencil brushes; **20** stencils; **21** ballpoint pen; **22** stencil card.

wine and food labels, giftwrap, greeting cards, wallpaper samples, photographs, travel magazines, and catalogs.

Gather together your collection of ideas and visualize them as decorated tiles. Think about the impact they will have collectively on the surrounding space. You do not have to limit yourself to the walls – also consider decorating the ceiling, floor, baseboard, and doors. Alternatively, simply work on a handful of tiles to give a hint of change. Think about color and the importance of light, and see how they can alter the effect of space within a room. The choice is infinite and is there to be tailored by you.

Tile decoration is non-utilitarian – neither lacquer nor varnish will give protection against persistent damp or abrasion. Therefore, decide carefully which tiles you want to decorate. Always work in a clean and dry environment. Wipe down tiles and grout with a surface cleaner prior to any paint or paper application. Allow adequate ventilation when spraying lacquer or using enamel paint as the fumes are extremely toxic and highly inflammable. A build-up of these fumes could prove dangerous.

Most important of all is that you enjoy embellishing your tiles. Confidence will come from trying the techniques and learning from your mistakes, so do not become despondent if your first attempts are unsuccessful. Designs can be easily removed before sealing, so push your imagination and experiment with ideas. Take pleasure from what you have created and feel proud when you achieve your goals.

Good sources of design material include photographs, cards, giftwrap, and wallpaper samples.

Special Paint Effects

The diversity of paint is limitless. An expanse of plain tiles presents an excellent face for decoration, and today there is a wide choice of tools designed specifically for special effects, although it is worth considering making your own. Paint can be applied using more or less anything – a crumpled plastic bag, a sponge, an old toothbrush, feathers, fabric, and so on. Try out different techniques on a few tiles and simply remove the paint before it dries. As glazed tiles are not absorbent, more durable paints such as enamel or acrylic are suitable. Make a sample board and test each color before applying it to a tile. Experiment and take into account the effect of light, which can create mood changes and completely alter the intended feel.

This checkerboard color wash design will add a fresh dimension to your kitchen or bathroom. Enamel paints are available in a wide choice of bright colors which, with their reflective quality, can create a feeling of playful modernity.

Checkerboard Color Wash

Equipment & Materials

Masking tape
Strips of cardboard
Sponge
Enamel paint (various colors)
Cotton swabs
Mineral spirits
Small paintbrush
White acrylic paint
Grout

1 Clean and prepare the tiles to be painted. Be sure to remove any dust particles and grease marks.

2

3

4

2 Mark tiles to remain white at random with masking tape. Those chosen can be changed as you proceed, but an initial indication is helpful. Place a strip of cardboard against the bottom edge of the tile you wish to paint, covering the grout. Using a sponge, dab paint along the cardboard edge. Placing it in the same way, move around the tile edges, working the paint toward the center.

3 Repeat this process for other colors, using a different piece of cardboard so that colors are not transferred to other tiles. Do not worry about smudging; this can be cleaned up later. Try to keep the work area free from dust and let it dry overnight.

4 Clean the grout, removing any excess enamel with a cotton swab dipped in mineral spirits. Do not let it run as this will leave a streak mark on the tile.

Stained grout can be touched up with white acrylic paint. Where it persists, press in a little fresh grout and remove any excess with a damp sponge.

SAFETY NOTE
Inhalation of fumes from enamel paint can be dangerous, so work in a well-ventilated area.

The Art Deco movement of the 1920s and 1930s was greatly influenced by Cubist art and the Industrial Revolution. The style is characterized by strong geometric lines, creating a confident and classical look. Inspiration for motifs such as the one illustrated can be found in the publications of the period.

Art Deco

Equipment & Materials

Selection of permanent ink pens
Paper
Ruler
Tracing paper
Masking tape
Transfer paper
Ballpoint pen
Cardboard
Cotton swabs
Lighter fluid

1 Measure a tile and draw it to scale on a piece of paper. Draw a line down the middle of the square, then halve each side again. Measure across to make 16 equal smaller squares.

2 Use a ruler and pen to draw your design. Start at the middle vertical line and work out using the ruled lines as a grid to achieve symmetry.

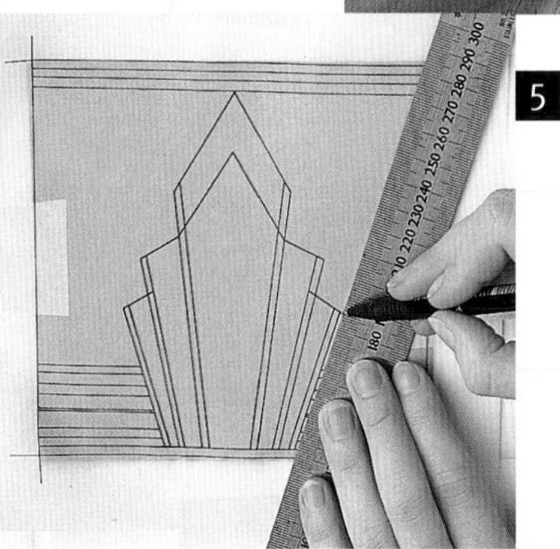
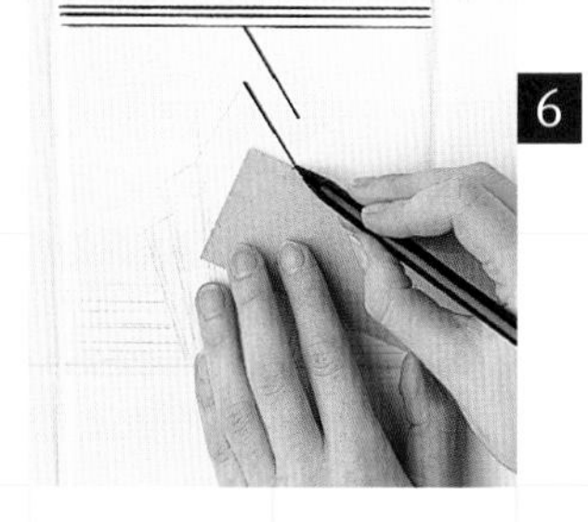

3 Experiment with color and try different combinations. If the design is to be used as a border, check that the image works as a repeat pattern. Keep the design as simple as possible; sometimes less detail means more impact. Use a black fine-point permanent ink pen to define the design accurately.

4 Place tracing paper over the design and hold in place with masking tape. Trace the corner edges of the tile and then the design itself. Take time to do this accurately; any mistakes will be carried through to your finished tiles.

5 Cut a piece of transfer paper slightly smaller than the tile and tape it to the tile, graphite side down. Position the tracing on top by lining up the corners, and tape it in place. Using a ruler and ballpoint pen, trace firmly and accurately over the *entire* design.

6 Remove the transfer and tracing papers. Retrace the design on the tile using a fine-point permanent ink pen. Use strips of cardboard to mark any lines – it is more absorbent than a metal or plastic ruler and therefore has less tendency to smudge. Wait a moment before lifting the cardboard directly away from the tile; do not pull away to the side as this will smudge the line.

7 Fill in block areas of color; work as closely to the outline as possible and then draw a neat black edge. Use a craft knife to scrape away mistakes. Repeat this process along the tiled wall. Remove smudges, fingerprints, and dust using a cotton swab soaked in lighter fluid.

Be adventurous with your patchwork design by varying patterns: mix flowers and stripes, dots and checks. Experiment with color – sometimes the more daring the clash, the more effective the combination. The simple techniques demonstrated here can be adapted to your own choice of designs.

Fabric Patchwork

Equipment & Materials

Fabric swatches
Window cleaner
Masking tape
Ruler and fine-point pen
Cotton swabs
Paper
Small piece of sponge
Scissors and removable adhesive
Toothpicks
Enamel or acrylic paint (various colors)
Small paintbrush
Thick cardboard
Spray-on clear gloss lacquer

1

1 Select designs from fabric swatches, gift wrap, or any printed material. Group the designs in batches of four, turn away for a few minutes and then look again. Rearrange the patches and repeat this process a couple of times. You will notice that certain fabrics work well side by side and other combinations are boring.

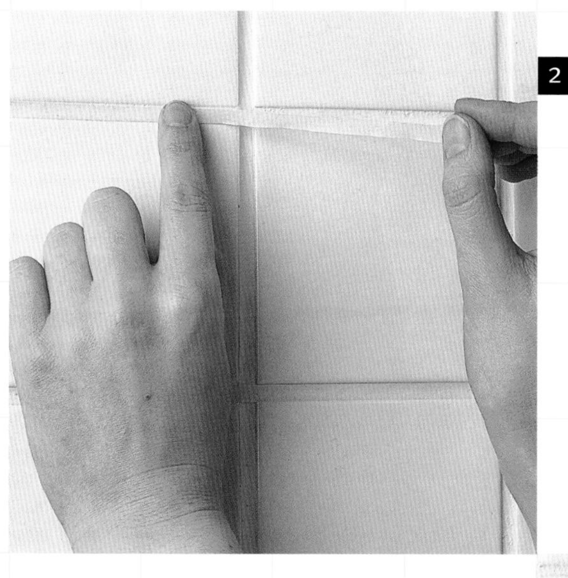

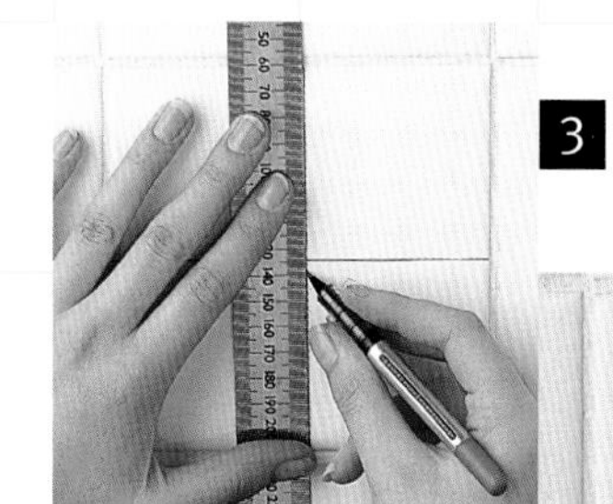

2 Before painting the tiles, remove dust and dirt using window cleaner. With fine strips of masking tape, cover the surrounding grout of tiles to be decorated. Run your finger along the groove to make sure the tape is firmly stuck down.

3 Measure the height and width of your tile and divide by two. Draw a pen line down the middle and across the center to make four equal squares.

4 Copy the first fabric swatch. For the "three-leaf clover" pattern shown here, the groups of dots were placed at random within the square. Although small motifs such as this do not need to be perfectly spaced, aim for some kind of rhythm.

5 Using a cotton swab dipped in water, wipe away the surrounding pen lines of that square only. Correct any smudges and let it dry. Cover with a sheet of paper for protection while working on adjacent squares.

6 To create a checked pattern, make a stamping block: cut a small square piece of sponge and use removable adhesive or glue to stick it to the end of a toothpick. Make one block for each color. Dab on a color at intervals in a vertical line, then fill in the gaps with a second color. Move to the second column and dab on a third color, starting one square down. Repeat this process until the entire square is full. Remove pen lines as before and when the paint is dry, cover with a sheet of paper.

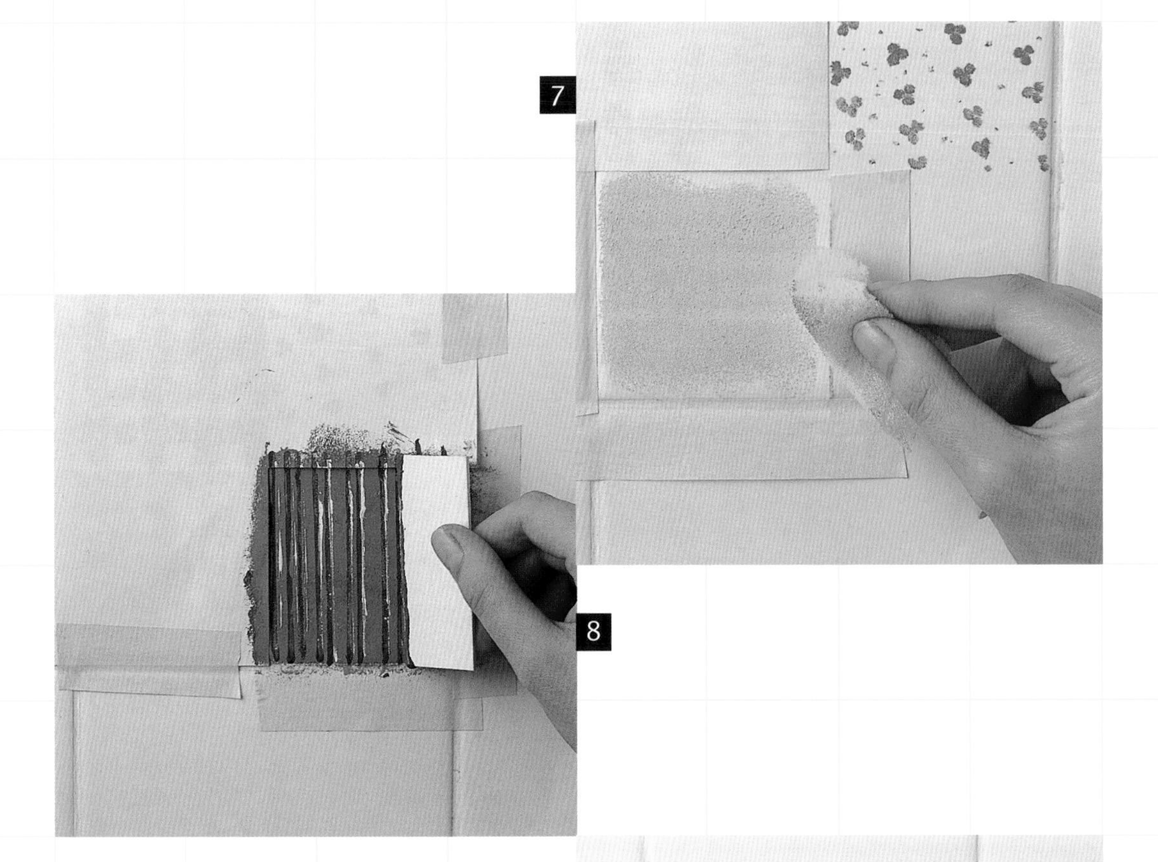

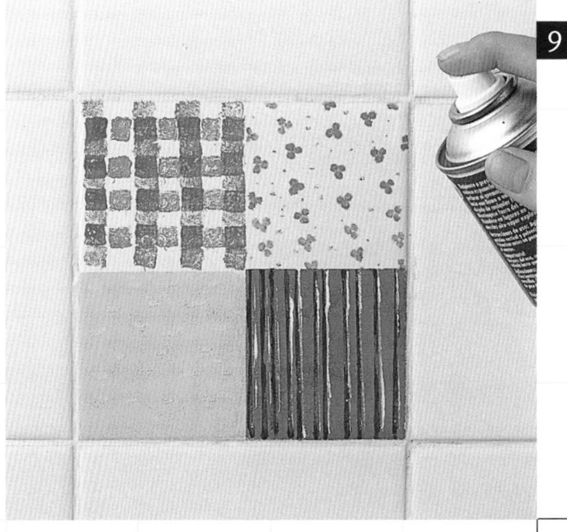

7 Apply masking tape to the right-hand edge of the third square. Dab on yellow paint using a sponge to create a textured surface. Place a few white dots at random, using a paintbrush, and sponge a second layer of yellow over the top to add detail. Let it dry, then protect with a sheet of paper.

8 On the fourth square, using a paintbrush, apply a layer of cornflower blue paint. Cut a piece of thick cardboard to the same depth as your square and coat the edge with purple paint. While the blue background is still tacky, dab on some vertical lines at random. Using another piece of cardboard, dab on some white lines. Repeat this process until the desired effect is achieved.

9 Remove all masking tape and wipe away any smudges. Make sure the surrounding tiles are free from grease marks and dust; if they are not, clean accordingly. When the paint is dry, spray on some gloss lacquer to protect the surface. Let it dry overnight.

SAFETY NOTE
Inhalation of fumes from enamel paint and spray-on lacquer can be dangerous, so work in a well-ventilated area. Take safety precautions as advised by the manufacturer or supplier.

DÉCOUPAGE

Découpage is the surface decoration of walls or objects using paper cutouts. It has been used for many centuries and is a novel way to transform tiles. Images can be extracted from maps, photographs, magazines, travel magazines, catalogs, sheet music, food and wine labels, and stamps. For a fun effect, look at children's coloring books, comics, greeting cards, candy wrappers, stickers, and gift wrap. Provided the paper is not too thick, the image can look very realistic. Spray-on glue is best for sticking down thin paper, such as newsprint and white craft glue can also be used: coat both sides of the cutout so it will dry flat. A photocopier, either color or black and white, is an invaluable device for découpage if a repeat pattern is required. Ask at your copy shop about special effects such as image stretching, elongation, and color manipulation.

This is a simple but effective way to add character to a room. The dado images used for this project were found in architectural reproduction catalogs, but you could draw them.

Architectural Feature

Equipment & Materials

Photocopies of image
Ruler
Craft knife
Cutting surface
Pencil
Spray-on adhesive
Spray-on clear gloss lacquer

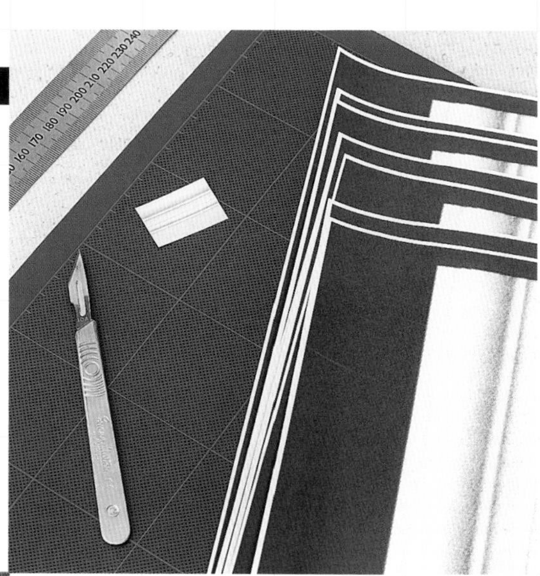

1 Make a color photocopy of the image, enlarged to tile height. (Although black and white copies cost less, color copies give a lot more definition.) Make as many photocopies as you need, allowing one image per tile.

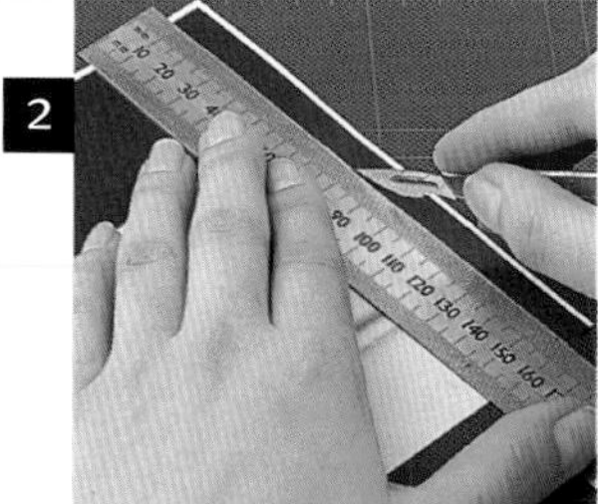

2 Use a ruler and craft knife, working on a cutting mat, to cut out the images as accurately as possible.

3

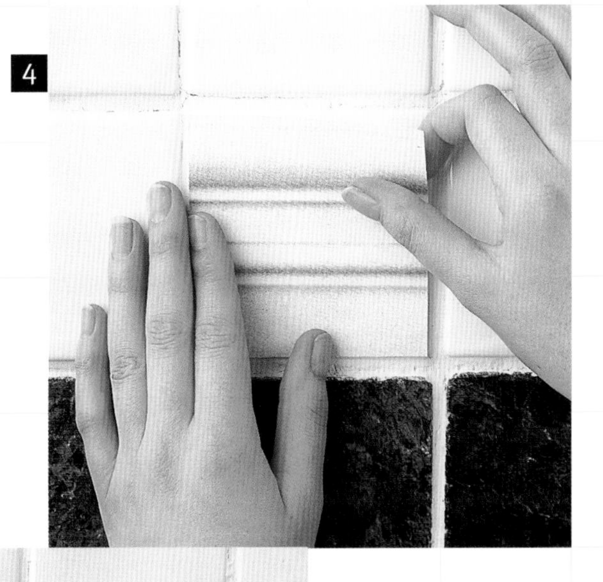

4

5

6

3 Place one of the photocopied images against a tile and, using a ruler and pencil, mark where it meets the grout. Work on one tile at a time since tiles may differ due to grout build-up and shrinkage. Remove the photocopy and, using a craft knife and ruler, trim according to the marks made.

4 Spray the back of the photocopy with an even layer of glue. Spray over newspaper so the surrounding area does not become tacky. Position it on the tile and smooth out any air bubbles.

5 Repeat this process along the row of tiles, checking the alignment of the images. If any of the images are reluctant to stick, gently peel away and reapply more glue. Trim any ragged edges, which will ruin the overall effect.

6 To add a little authenticity, place a ruler at the point where the shaded horizontal line, approximately three-quarters of the way down the dado image, graduates into white, and cut across with a craft knife. Cut consistently through the images on the other tiles. Remove the lower part of the image to expose the gleaming white tile underneath.

7 Working from left to right, spray the dado découpage with lacquer. Apply another layer, first waiting a couple of minutes to avoid making drip marks. Let it dry overnight, keeping the area dust-free and well ventilated.

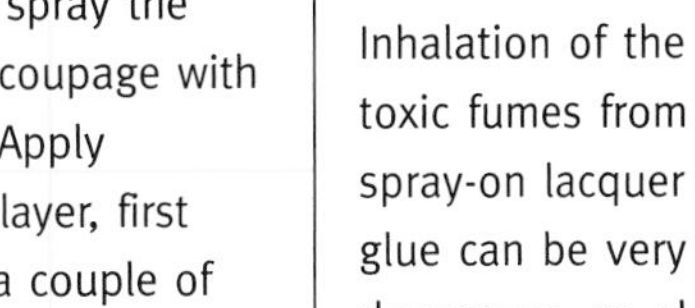

SAFETY NOTE
Inhalation of the toxic fumes from spray-on lacquer and glue can be very dangerous, so always work in a well-ventilated area. Follow the manufacturer's recommended safety procedures when using these products.

This sun-drenched view will give a whole new dimension to your tiles. Peep through the grapevine and let your imagination do the rest. Take advantage of this lustrous palette and capture the very essence of the sunny south.

Mediterranean View

Equipment & Materials

Color photocopies
Cutting mat
Ruler
Craft knife
Scissors
Pencil
Spray-on glue
Spray-on clear gloss lacquer

1 Photocopy the Mediterranean view, enlarged to cover the width of two tiles. Photocopy enough border repeats to go around the edge twice to allow for any overlaps and filling in of gaps. Cut out on a firm cutting mat. Use a ruler and craft knife for straight edges and scissors for detailed edges.

2

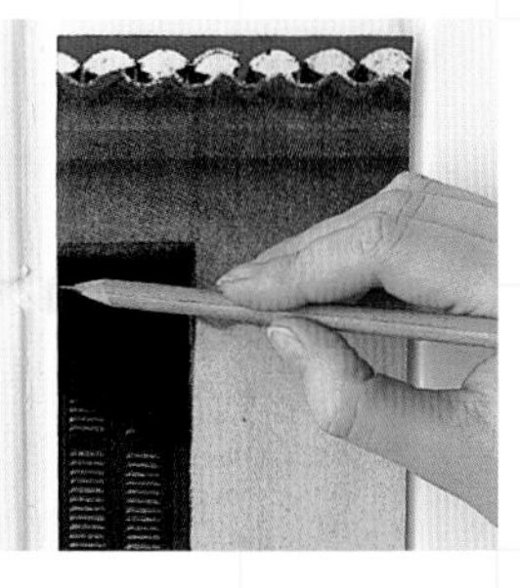

5

2 Measure and mark the midpoints of the view at the top and bottom. Cut into two pieces.

3 Place one strip against the tiles and make a pencil mark where it meets the grout. Cut along this edge. Repeat this process of positioning, marking and cutting, tile by tile.

4 Spray the reverse side of each section of the view with glue, one at a time, and place in position. Trim away any overlaps caused by uneven tiles using a ruler and craft knife. Stand back to check that the view is straight and that the pattern created by the grout complements the image.

5 Position the grapevine border around the view. Start in the top left-hand corner and decide in which position it works best. Mark with a pencil where the grout lines occur and cut out. Work your way around the border at random, hiding edges and filling in gaps.

6 Check that the surrounding tiles are clean and dust-free, and spray with two coats of lacquer. Wait for a couple of minutes between each coat to avoid making drip marks. Keep the area dust-free and leave to dry overnight.

SAFETY NOTE
Inhalation of fumes from spray-on lacquer and glue can be dangerous, so work in a well-ventilated area. Follow the manufacturer's instructions regarding safety, and store well out of the reach of children.

The subtle, glowing colors of fruit look particularly striking against a sparkling white background. You can choose just one image to create a decorative focal point, or use several to scatter across a tiled wall.

Fruit Motif

Equipment & Materials

Four plain glazed tiles
Photocopies of image
Pencil
Ruler
Craft knife
Cutting surface
Spray-on glue
Small sponge
Spray-on clear gloss lacquer

1

1 Make sure the tiled surface is completely clean and dry. Place the image at the center point of the tiles and decide whether you want it to be enlarged or reduced in size.

2

3

4

5

2 When you have the correctly sized image, center it over the tiles, and mark where the position of the vertical and horizontal lines of grouting meet the paper. Lightly draw a pencil line between these marks so that you divide the image into quarters.

3 Cut the image into quarters along the ruled lines, then carefully cut out the picture with a craft knife, working around the details for a professional-looking finish. Keep the paper "frame" as you will use this as a template to position the image on the tiles.

4 Dealing with each quarter in turn, glue the back of the cutout picture, line up each quartered frame with the grouting lines, and use it to position each part of the picture in the right place. Press the design from the center out to remove any air bubbles.

5 Blot any excess glue from around the design using a small sponge squeezed out in warm water, and when the glue has had time to dry completely, spray the découpage evenly with lacquer. Keep the area dust-free and leave to dry overnight.

SAFETY NOTE
Inhalation of fumes from spray-on lacquer and glue can be dangerous, so always work in a well-ventilated area. Follow safety precautions as advised by the manufacturer or supplier.

Bring life to your tiles with a real garden trellis, adorned with an array of flowers from cards, gift wrap, and gardening magazines. An enchanting concept with visual charm, it can replace pictures or simply add character to a room.

Real Garden Trellis

Equipment & Materials

- *Garden trellis*
- *Saw and sandpaper*
- *Acrylic paint*
- *Paintbrush*
- *Pictures of flowers*
- *Photocopies*
- *Cutting mat*
- *Craft knife*
- *Masking tape and removable adhesive*
- *Pencil and ruler*
- *Spray-on glue*
- *Spray-on clear gloss lacquer*
- *Drill*
- *4 screws*

1 Cut the trellis with a saw to the size required. Smooth rough edges with sandpaper, and wipe with a damp cloth to remove any dust. Apply the paint with a brush and let it dry. Paint a second coat if necessary.

2 Make two color copies of the flowers to be used. Working on a cutting mat, use a new craft knife blade to carefully cut out the flowers, keeping as much detail as possible to maintain the realism.

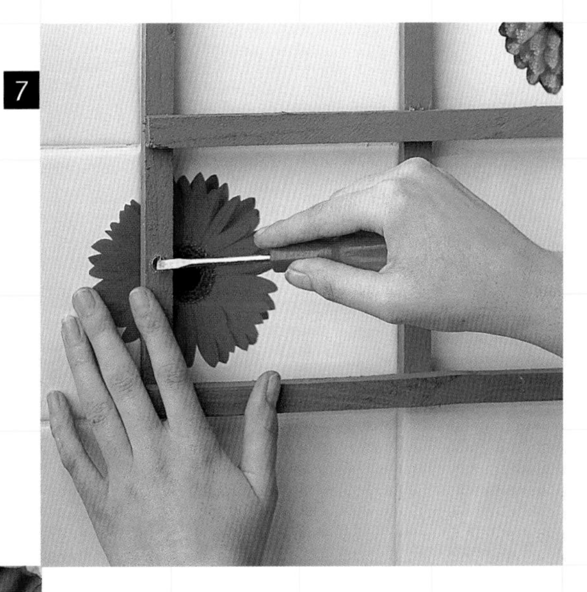

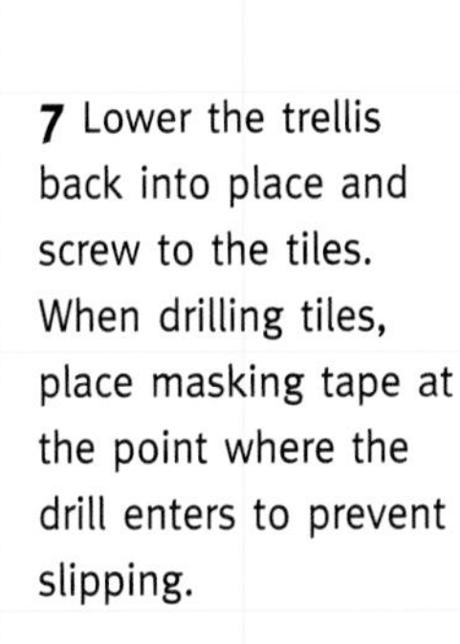

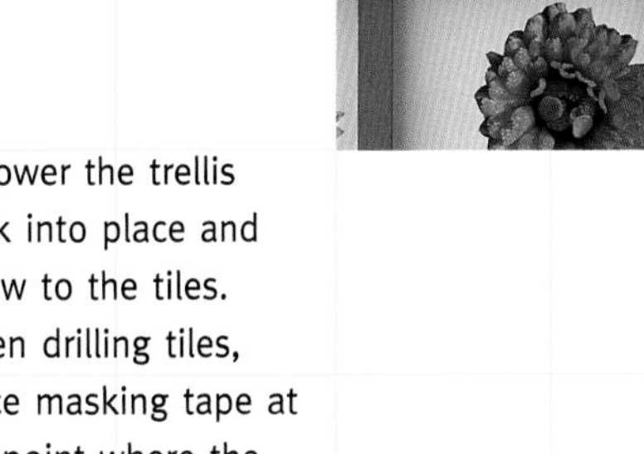

3 Position the trellis on the tiles. If your tiles are approximately the same size as the trellis grid, cover as much of the grout line as possible. If they are not, place the trellis halfway between the grout lines. Hold the trellis in place using masking tape: thread a piece through each square along the top row to form a line of loops and then stick each to the tile with a horizontal strip of tape.

4 Use the horizontal and vertical lines of the trellis as a guide to position the flowers. Only place flowers which have a duplicate under the wooden grid. Balance the pattern to maintain a loose, scattered feeling. Use removable adhesive for repositioning the flowers.

5 Flip the trellis up and away from the tiles and tape it back – do not detach it from the original tape loops, as repositioning may be difficult. Note the position of the flowers with pencil marks. One by one, remove the flowers, spray with glue, and stick in position. If the grout lines are not covered by the trellis, divide the flowers, using a craft knife and ruler, where the grout falls.

6 Make sure the surrounding tiles are free from grease marks and dust, and that the flowers have been well stuck down with no ragged edges. Spray with a layer of lacquer and leave to dry.

7 Lower the trellis back into place and screw to the tiles. When drilling tiles, place masking tape at the point where the drill enters to prevent slipping.

8 Take the doubles of those flowers obstructed by the trellis. Position each duplicate on the trellis, aligning them with the original flowers. Mark the area hidden by the trellis with a pencil and cut out using a ruler and craft knife.

9 Spray glue on the reverse of each flower section and stick them to the trellis, aligning petals and trimming overlaps.

10 With the main pattern complete, enhance with several flowers. Try sticking them on the trellis itself, on the tiles, or positioning them outside the trellis to give unity to the wall.

11 Check that the tiles are clean and free from dust, and spray with lacquer. Wait for a couple of minutes before spraying again, to avoid making drip marks. Spray in and around the trellis to make sure all découpaged surfaces are covered. Let it dry overnight, keeping the area dust-free.

SAFETY NOTE
Inhalation of fumes from spray-on lacquer and glue can be dangerous, so work in a well-ventilated area. Always wear goggles to protect your eyes when drilling tiles.

During the eighteenth century, print rooms were fashionable as personal and intimate rooms for ladies living in Georgian houses. Use photocopies of old engravings or prints, edged with decorative borders, tassels, and ribbons, to create the grandeur reminiscent of the Georgian style.

Eighteenth-century Print Room

Equipment & Materials

Photocopies of prints
Ribbon and braid
Needle and thread
Cutting mat and ruler
Craft knife and scissors
Pencil and spray-on glue
Removable adhesive
Spray-on clear gloss lacquer

1 Photocopy the print and its frame. For a sharp image, copy them separately; the glass will cause light reflection. Choose pieces of ribbon and braid, and photocopy. (These can be placed in the empty frame.)

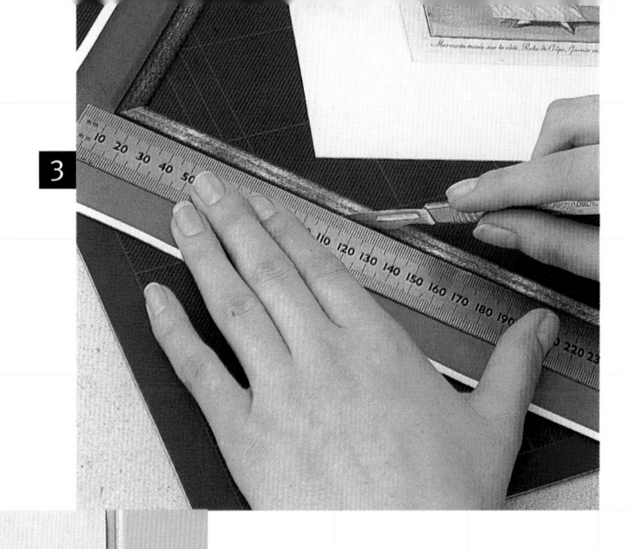

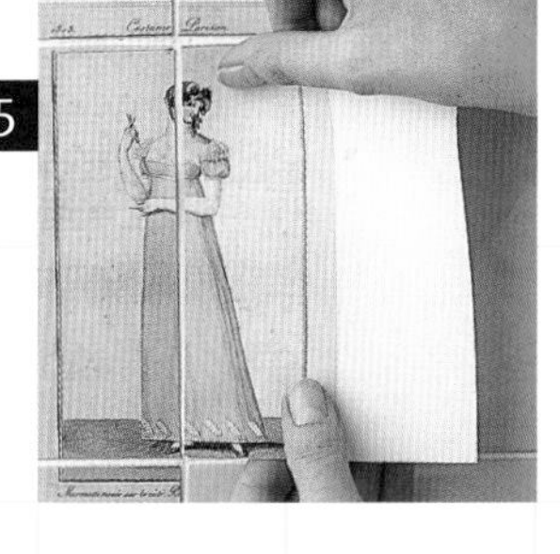

2 Now wrap the braid into a coiled circle. Using a needle and thread, baste it in position on one edge only and photocopy again.

3 On a cutting mat, use a ruler and craft knife to cut out the photocopied images. Be sure to cut the frame edges straight, and remember to cut around the mat and not just the print. Cut a "V" shape at each end of the ribbon. Use scissors to shape the coiled braid into a neat circle.

4 Cut the print down the middle and place one half against the tiles. Make a pencil mark on the copy where it meets the first horizontal line of grout. Cut along this line. Repeat at the next horizontal lines and then with the other half of the print until the whole image has been divided.

5 Spray glue on the back of the first print section. Spray onto newspaper to protect the surrounding area from becoming tacky. Position the print section on the first tile and smooth away any lumps or air bubbles with your fingers. Check that the edges are firmly stuck. Repeat the process for the other print sections.

6 Place the ribbon diagonally toward the center at the lower edge of the print mat and mark where they meet. Cut along this line. Repeat with a slightly longer piece of ribbon on the other side. Spray the pieces with glue and stick in position.

7

8

9

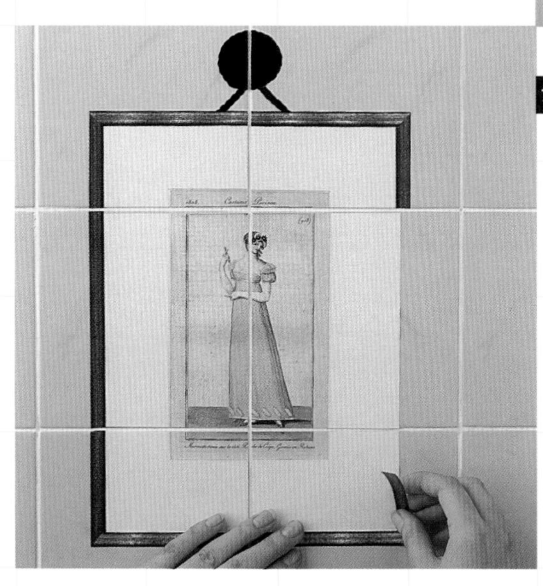

10

SAFETY NOTE
Inhalation of fumes from spray-on lacquer and glue can be dangerous, so work in a well-ventilated area. Take safety precautions as advised by the manufacturer or supplier.

7 Cut two short lengths of straight braid and place along the middle top edge of the mat at a 45-degree angle to each other. Spray them with glue and stick in position. Divide the coiled braid in half, spray with glue, and position on each side of the grout slightly overlapping the straight lengths of braid.

8 Use removable adhesive to hold the frame in position. Check that where the mat meets any ribbon or braid, it is hidden by the frame and that everything looks well balanced.

9 Starting at the middle bottom edge of the mat, make a pencil mark where the frame covers the grout. Trim out the pieces with a craft knife.

10 Spray the frame pieces with glue and position carefully. The frame holds the whole image together so take extra care at this stage. Smooth away any air bubbles and check that the edges are firmly stuck to the tile.

11 Check that the découpage and surrounding tiles are free from grease marks and dust, then spray with lacquer. Wait for a couple of minutes to avoid making any drip marks, then spray again. Keep the surrounding area as dust-free as possible, and let it dry.

STENCILING

Stenciling has become increasingly popular, with an impressive choice of ready-made stencils available today. However, nothing competes with the versatility of developing original designs to suit your own individual style. The scale of a stencil can be changed to any size required, and in other ways can be tailored to specific requirements. Cardboard, a sharp blade, and patience are the key tools to successful stencil-making. Stencil card (oiled manila card) is highly recommended because it is tough – it will withstand repetitive use – and it resists the moisture in paint. Regular cardboard is absorbent and makes a weak stencil, prone to warping or tearing. A cutting mat allows more control when you are cutting out.

The Greek key border motif originated from ancient Greek architecture. It is a maze of right-angled lines that fit together to make a simple but effective repeat pattern.

Greek Key Border Motif

Equipment & Materials

Ruler
Pencil
Stencil card
Cutting mat
Craft knife
Double-sided tape
Acrylic paint (in color of choice)
Sponge or stencil brush
Cotton swabs
Spray-on acrylic gloss varnish

1 Measure and draw your tile to size with a pencil on a piece of stencil card, allowing a 1-inch (2.5-cm) border all the way around. Divide into 8 equal strips vertically and 7 strips horizontally. Mark an arrow in the top left-hand corner to indicate which is the top.

1

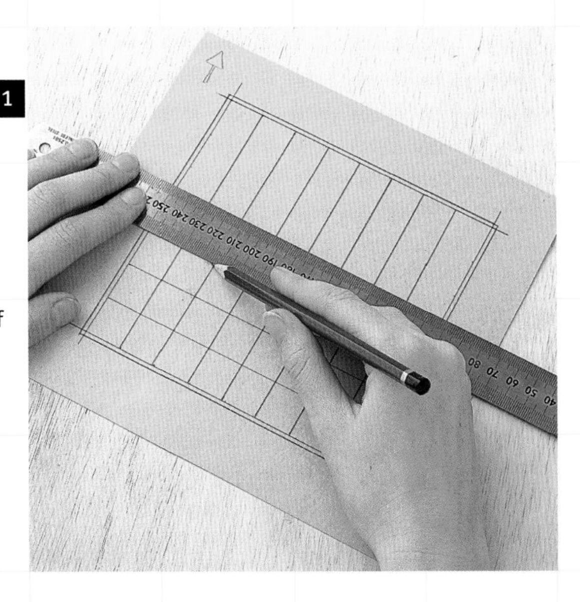

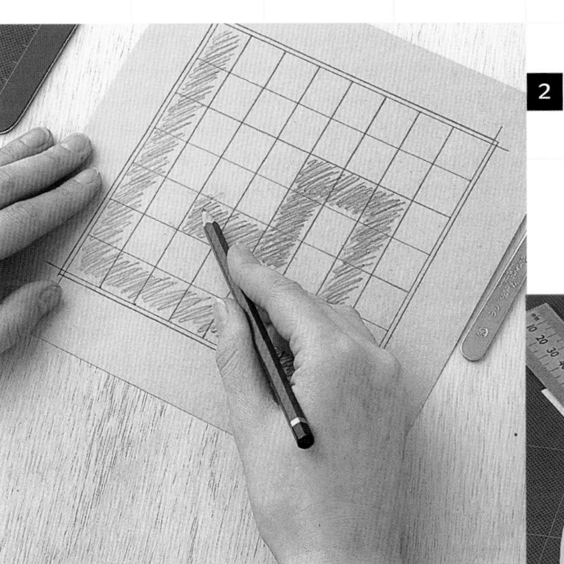

2

4

3

2 Starting at the left-hand side of the grid, shade 7 squares down and 6 across (i.e. to the second square in from the right-hand edge). Shade 4 up, 2 across to the left, 2 down, 2 across to the left, 4 up, and 5 across to the right. You should have used the whole length and breadth of the grid to create a Greek key pattern.

3 Working on a cutting mat, use a craft knife and ruler to cut out the shaded area. Make sure the lines are straight and cut at right angles.

4 On the reverse side of the card, place strips of double-sided tape along the motif edges. Use ¼-inch (5–10mm) wide tape – if it is any wider, the stencil could tear when being removed from the tile. Trim any overlaps using a craft knife and ruler.

5 Peel away the backing from the tape and position the stencil on the tile. Align the bottom edge of the motif with the bottom edge of the tile and check that it is straight. To realign it, run a water-soluble pen along the inside bottom edge and gently peel back the stencil. Remove the pen and reposition the stencil. Dab on the acrylic paint using a sponge or stencil brush. (Make sure the consistency of the paint is not too thin.)

6 Gently peel away the stencil, dislodging the central area first to avoid tearing. Clean up any smudges: scrape them with a craft knife and wipe away with a cotton swab dipped in water. When the paint is dry, repeat the process on the next tile.

7 Check that the tiles are free from fingerprints and dust, and spray from left to right with acrylic lacquer. Wait for a couple of minutes before applying another coat, to avoid making drip marks. Let it dry.

5

6

SAFETY NOTE

Inhalation of fumes from spray-on acrylic varnish can be dangerous, so work in a well-ventilated area. Take safety precautions as advised by the manufacturer or supplier.

At first glance, a stenciled design may seem complicated to do. However, when it is broken down into stages, is easier than first anticipated. The "seashore" is a versatile theme used in the kitchen or bathroom, or a child's playroom. It is fresh and fun with an unlimited source of forms to choose from.

By the Seashore

Equipment & Materials

Ruler
Pencil
Tracing paper
Cutting mat
Craft knife
Pictures of seashore objects
Masking tape
Paper
Stencil card
Spray-on glue
Sponges
Acrylic paint (in color of choice)
Spray-on clear gloss lacquer

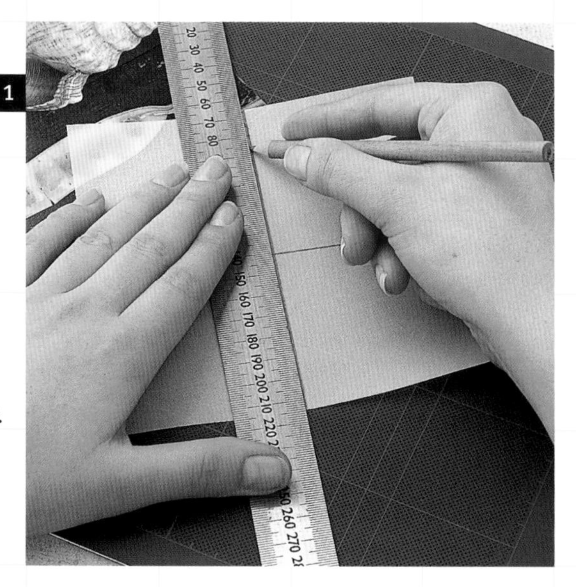

1 Measure your tile and with a pencil draw it to size on a piece of tracing paper. Cut it out on a cutting mat, using a craft knife and a ruler. Draw a line down the middle and across the center to make four equal squares.

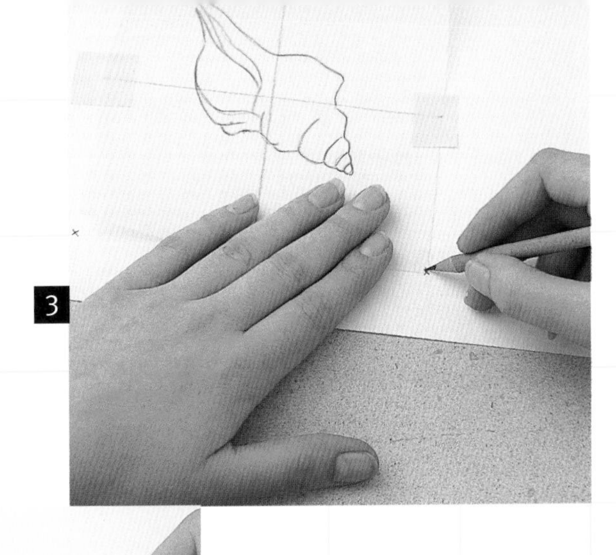

2 Position your chosen image near the center of the square and accurately trace as much detail as possible.

3 Tape the tracing onto some blank paper. If you use very thin paper, put something soft underneath such as a piece of cardboard or an old magazine. Mark the corners of the tile on the paper so you will know if the tracing has moved and where to cut the tile perimeter. Retrace the image, pressing hard with your pencil.

4 Remove the tracing and fill in the lines of the imprint on the paper. If the lines are too faint, shade in the area using the side of your pencil. This tracing technique allows for your imagination to exaggerate curves and simplify detail.

5 Divide your drawing into positive and negative parts. The positive parts are the areas of your design that will be painted and the negative areas are those that will be left blank. If an area is to be blank it will remain covered, therefore the parts of the stencil to be left intact are these negative areas. Cut out the positive shapes. Every now and again, stop to check the shapes you are making by holding the paper against a colored surface.

6 When you are satisfied with your design, place it on top of a piece of stencil card. Mark the corners and cut to your tile size. Tape the cutout drawing in position on the card and draw around the inside edge of the template using a sharp pencil. Shade these areas to indicate they are to be cut out.

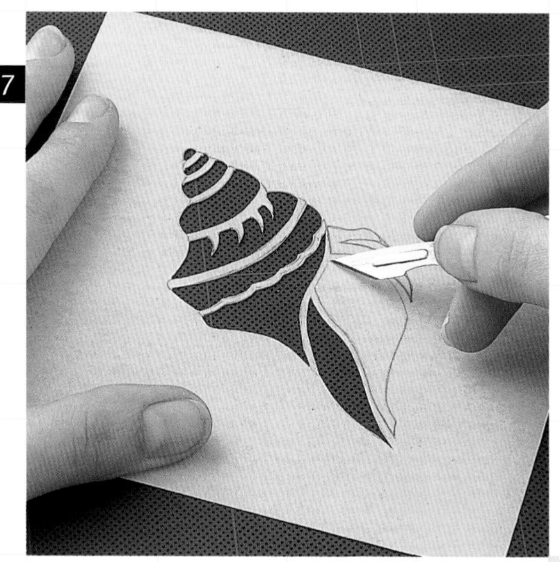

7

7 Remove the drawing and, using a new craft knife blade, cut around the shaded areas on the card. Rotate your stencil as you work so your cutting action is always in the same direction. Relax your hand to go with the curves, but be firm to achieve a clean cut line.

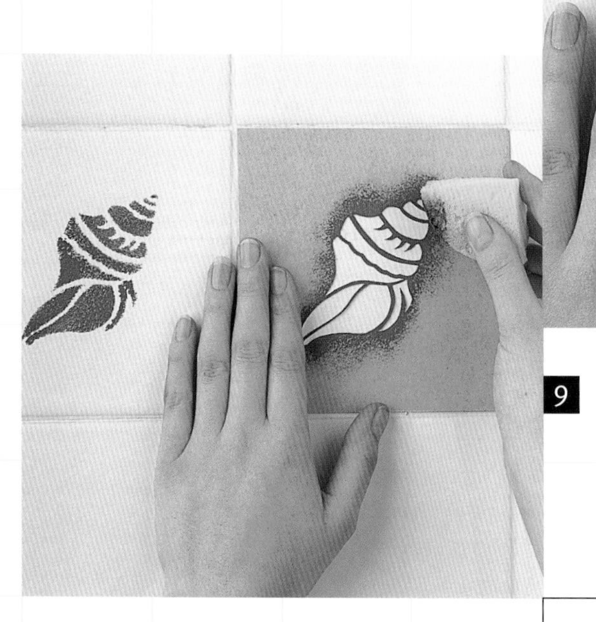

8

9

8 Spray the back of the stencil with a little glue to stop it from slipping and position it against a tile. Using a small piece of sponge, dab on the paint. Work the sponge gently to avoid paint squeezing out and running underneath the stencil edges.

9 After painting, remove the stencil directly away from the tile to avoid smudging. Make sure there is no paint on the reverse side of the stencil that will mark the next tile and continue working along a row of tiles. Check each time that the stencil is straight and the image well centered.

10 When the paint is dry, make sure the surrounding tiles are clean and free from dust, and spray with lacquer. Wait for a couple of minutes before applying another coat, to avoid making drip marks. Leave to dry overnight, keeping the area dust-free.

SAFETY NOTE

Inhalation of fumes from spray-on lacquer and glue can be dangerous, so work in a well-ventilated area. Take safety precautions as advised by the manufacturer or supplier.

TRANSFER DESIGN

Transfer paper is an invaluable innovation for copying impressive patterns with intricate detail onto a variety of surfaces. Available from most good artist's supply stores, it enables you to reproduce a work of art with the precision of the original. From Dutch Delftware to William Morris designs, any image that can be traced can be transferred. Transfer paper is a thin tissue paper with graphite backing. It is placed under a design and traced over to deliver an identical line of graphite, true to the original. This is durable enough to withstand smudging, but can be removed with a damp cotton swab when no longer needed. Use acetate to trace complex lines as tracing paper can obscure details.

Tracing and transfer paper work together well. Creating a design on tracing paper allows the freedom to experiment and any mistakes made can be remedied. Transfer paper can be used to make an identical copy, down to the finest of detail, onto most hard surfaces. This is an opportunity to use your drawing skills to accurately transform your design into an impressive painting.

Potted Orange Tree

Equipment & Materials

- *Tracing paper*
- *Ruler and pencil*
- *Masking tape*
- *Transfer paper and pen*
- *Selection of acrylic paints*
- *Fine paintbrushes*
- *Toothpicks*
- *Cotton swabs*
- *Spray-on clear gloss lacquer*

1 Cut a piece of tracing paper to the size of the tiled area to be decorated. With a ruler and pencil, divide the paper widthwise into three. Draw a vertical line down the middle. Tape into position over the tiles.

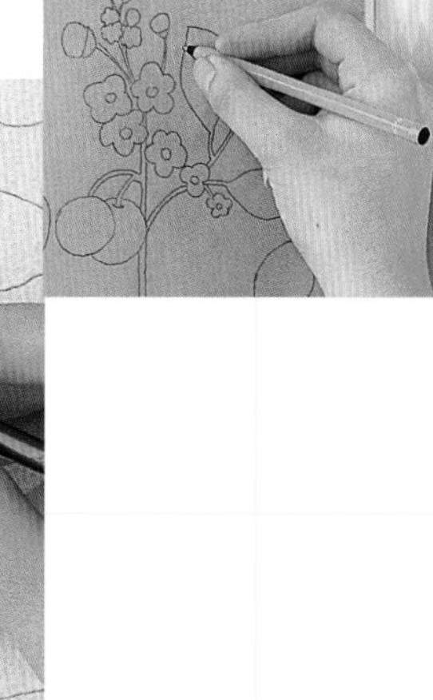

2 Freely draw a plant pot in the bottom section. Relax your hand and, using the middle line as a guide, continue to draw the tree section by section, making meandering curves to simulate the stems. Draw them at random, decreasing in size as you near the top. Draw around the lid of a small jar to make the oranges.

3 Place a sheet of transfer paper underneath the tracing with the graphite side toward the tiles, and tape into place. Press firmly with a ballpoint pen and trace. You do not need a new piece of transfer paper per section; simply slide the same piece down.

4 Remove the papers and start by painting the oranges. Vary the color tone, adding white specks to give the dimpled texture of orange peel. Mix pale lemon with hard yellow to make fresh summer tones to create the flowers.

5 When the paint is dry, use a toothpick to scrape away any hard edges to soften the contours of the flowers and fruit. Remove any unwanted graphite using a cotton swab dipped in water.

6 To paint the leaves, mix together a rich palette of purples and blues. Spontaneously apply colors. Keep the contours flowing by sculpting hard edges into sweeping curves.

7 Paint the stems olive green. Start at the top and work down to avoid smudging. Use a cotton swab to dab white paint into the center of the flowers, followed by small olive dots.

8 Use a fine-pointed brush and olive paint to draw the central vein of the leaves. Let it dry.

9 Check that the surrounding tiles are clean. Remove marks using a cotton swab soaked in lighter fluid. Spray with lacquer, waiting for 2 minutes before applying another coat, to avoid drip marks. Let it dry overnight and keep free from dust.

SAFETY NOTE

Inhalation of fumes from spray-on lacquer can be dangerous, so work in a well-ventilated area. Take safety precautions as advised by the manufacturer or supplier.

Delftware designs originated in Holland in the seventeenth and eighteenth centuries and depict the everyday scenes of Dutch life. Charming characters are painted with bold brushstrokes, loosely detailed by tone. The Delft style gives an uplifting and refreshing look to tiles.

Delftware

Equipment & Materials

Delftware source material
Ruler
Acetate
Masking tape
Permanent ink pen
Transfer paper
Ballpoint pen
Fine paintbrush
Acrylic paint (in colors of choice)
Cotton swabs
Spray-on clear gloss lacquer

1 Sources for the Delft style include postcards, pottery, and reference books. Photocopy and enlarge Delft images to tile size, preferably using a color copier for greater tone definition.

2

3

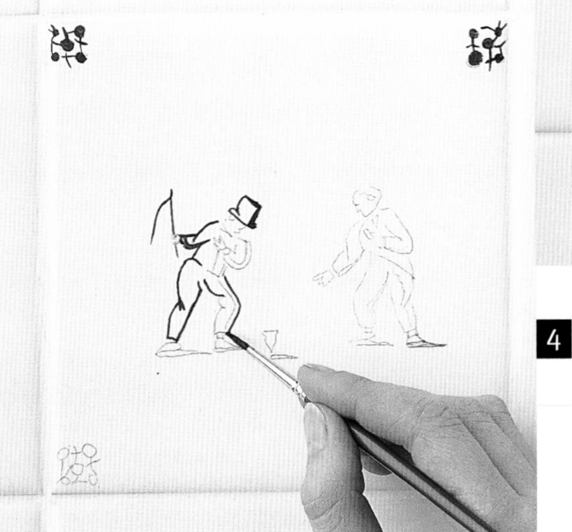

4

2 Cut a piece of acetate to the tile size, place it on top of the photocopy, and hold in place using tape. Carefully trace over the image using a permanent ink pen.

3 Position the acetate on the tile and hold in place with tape. Slide a piece of transfer paper underneath with the graphite side toward the tile and tape in place. Use a ballpoint pen and press firmly to accurately retrace the image.

4 Remove the acetate and transfer paper. Use a fine paintbrush to apply paint over the lines. Relax your hand and try to work as freely as possible to recreate the nature of this style. Remove any unwanted graphite with a cotton swab dipped in water.

5 Check that the tiles are free from fingerprints and dust, and spray with lacquer. Apply 4–5 coats allowing a couple of minutes between each application to avoid making drip marks.

SAFETY NOTE
Inhalation of fumes from spray-on lacquer can be dangerous, so work in a well-ventilated area. Take safety precautions as advised by the manufacturer or supplier.

The tiles shown on these pages were made and decorated by contemporary designers, who were inspired by both modern and traditional themes. All demonstrate the almost endless choice of colors, textures, and styles that can be used to create stunning effects.

Gallery of Tiles

▶ BRONWYN WILLIAMS-ELLIS

Triptych
A private commission in which molded and hollowed out earthenware slabs make a frame for a "window" of hand-made clay tiles. They are hung on the wall, not stuck with adhesive. The design was created with slips, stencils, and cuerda seca (dry line), with handcarved border patterns, roulettes, and stamps. Very little glaze was used, giving a chalky, matte appearance.

◀ MARLBOROUGH CERAMIC TILES

Tulip and plate panels
These naturalistic tulips in their blue and white vase are designed to be set into plain tiles as a decorative element to a kitchen wall. The white base tiles are hand stenciled in a modern interpretation of a Victorian style.

▶ DECORUM CERAMIC STUDIO

"Harriet" series
These hand decorated tiles were designed to complement the fashionable dragged look, popular in many homes today. Each tile is painted over glaze and then dragged with a natural bristle brush. A large choice of colors in the palette makes an extremely flexible range that can be mixed and matched, and, as in this case, laid at different angles.

▲ MARY ROSE YOUNG

Metallic tile
Bright colors on a dark background, raised surfaces, and metallic touches give an ultra modern appearance. This tile is handmade earthenware, with decorations built on to the overglazed color and biscuit fired. A final luster firing produces the gold and silver quality. Because of the uneven surface, this type of tile is not suitable for areas that need a lot of cleaning.

◀ MARY ROSE YOUNG

Ship, heart, fish, and flowers
These bright, childlike images in cheerful primary colors are ideal for wall tiles in showers, bathrooms, or kitchens. They evoke a vacation mood, and combine the simple lines of Dutch picture tiles with uninhibited color well suited to modern homes. The designs are painted on industrial biscuit tiles, using underglaze pigments with a transparent glaze on top to provide an impervious, washable surface.

MAKING TILES

MAKE YOUR OWN TILES, BY USING VARIOUS METHODS AND WORKING WITH DIFFERENT TYPES OF CLAY. ALL THESE PROJECTS HAVE DECORATIVE FINISHES THAT REMAIN PERMANENT AFTER FIRING IN A KILN.

At the beginning of each ceramic project in this book, all the required tools are listed in the order of their use. A few are specialized pottery tools, which must be purchased from pottery craft suppliers, but many are general household or workshop items.

Tools and Equipment

Apart from a kiln (see pages 56–9), the most expensive piece of equipment required for any particular project is the clay extruder. The two extrusion projects (see pages 84 and 86) demonstrate a partly mechanized form of tile production, but the same result can be achieved by hand rolling and cutting the clay. The purchase of such major items is not necessary for the home hobbyist since local potteries and college workshops can generally provide facilities.

1 Sifter.

Sifters

If you intend to mix your own glazes, a set of sifters is essential. These come in a variety of mesh sizes which are described by the number of wires per inch in the weave, i.e., 60 mesh, 100 mesh. They are available in various diameters and are used with coarse brushes in appropriate sizes for working on different quantities of material. For large quantities of material, rotary sifters can be purchased with an integral set of three brushes and a supporting aluminum frame.

Slip Trailers

These are used for trailing liquid clay (slip) in decorative patterns onto a wet clay surface, in much the same way as piping icing onto a cake. There are various types, but all have a collapsible pouch into which the slip is poured and a narrow nozzle, either of glass, rubber or plastic, for drawing the design.

Potter's "Kidney"

This is a piece of rubber or metal (as the name suggests, shaped like a kidney) used for smoothing down clay and pushing any granular material below the surface. Kidneys are made in a number of sizes and with varying flexibility.

Potter's Needle

Consisting of a narrow steel point set in a small wooden handle, a needle can be used for several purposes, such as marking dimensions on a slab of clay, gauging the depth of a glaze, or pricking air bubbles in a wet surface.

Cutting Equipment

A variety of cutting tools are used throughout the projects.

For cutting out clay shapes and making cardboard or plastic templates, a steel craft knife or similar sharp blade ensures accuracy and gives the cleanest edge. An even easier tool to use for making clay shapes is a tile cutter. This consists of a galvanized or stainless steel die, like a pastry cutter, inside of which is mounted a spring-loaded ejection plate. It is simply pushed into the clay slab and will hold the cut piece until it is ejected.

When modeling clay, one of the most invaluable tools is a simple plastic blade for paring down edges. It costs almost nothing to make and is easily cut from a thin piece of rigid plexiglass. Make sure that the working edge is as smooth and clean as possible by rubbing it with fine sandpaper.

For making a linoleum printing block (see page 112) there are purpose-made tools both for cutting lines and gouging out larger areas. Some have interchangeable nibs.

A potter's clay-cutting wire, which is simply a twisted wire that can be pulled taut between two wooden toggles, is useful for slicing clay into workable portions. For more precise cutting, a potter's harp, consisting of fine wire stretched across a steel frame, gives a cleaner line.

Workshop Tools

Although some basic woodwork and metalwork skills are needed to make the customized frames and dies for particular projects, this work can be carried out fairly inexpensively by a local carpenter or metal shop. For those with some craft experience, the most important items are an electric drill, a wood plane, a set of hand saws, and a jigsaw.

Mold Boards

A set of adjustable mold boards is essential for the cast tile projects (see pages 78 and 94). These can be made from planed wood or from laminated blockboard that is sold for shelving. Four boards are needed, and to be usable for all the projects, each must be 6in (150mm) deep and at least 28in (700mm) long. One end of each board must be cut perfectly square to its bottom edge. Onto the top edge of each of these squared ends, an "L" shaped metal bracket is screwed so that it protrudes beyond the end by the exact thickness of the boards. This bracket will hook over the side of its adjacent board and link all four into a square. Although each bracket fits snugly over the top edge of its neighboring board, there should be just enough slack to allow the dimensions of the square to be adjusted. These boards form the confining walls of the plaster mold during its casting, and when they have been set to the correct size for a particular model, they are easily secured in place with lumps of clay on their outside faces.

Dipping Frames

Two simple tools for lifting and dipping tiles can be formed from a wire coat-hanger. The first, for dipping tiles in wax emulsion, is made by cutting off one wing of the hanger as close to the hook as possible and then bending the resulting wire "V" in half to form a right-angle cradle. This is accomplished by clamping the open ends in a vise, and using a pair of pliers and a hammer to bend the wire.

The second tool, used for dipping very thin tiles into a glaze, can be made from the other wing of the coat hanger. This should also be cut off tightly to the hook, then all you need to do is bend over the last fraction of an inch of the open ends to form small hooks.

Household Tools

Many of the projects use general kitchen items: a rolling pin, pitchers, bowls, an icing side-scraper. A set of scales is also useful. Once such items have been used for pottery, they should not be used for any other purpose.

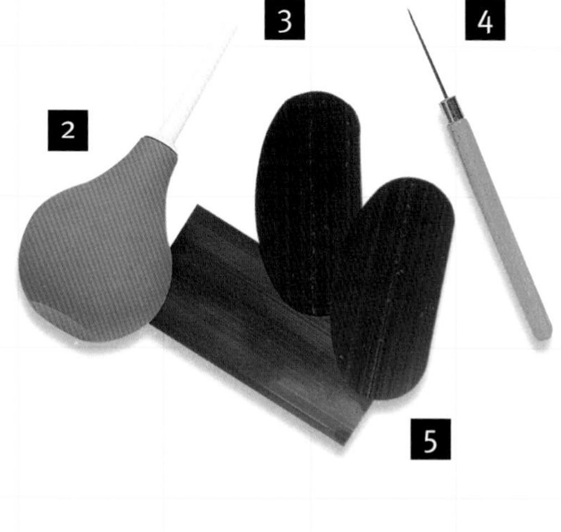

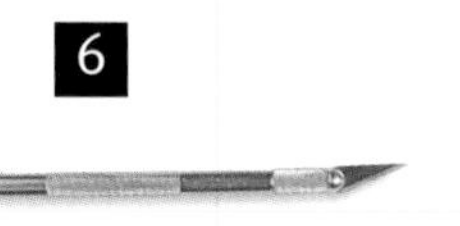

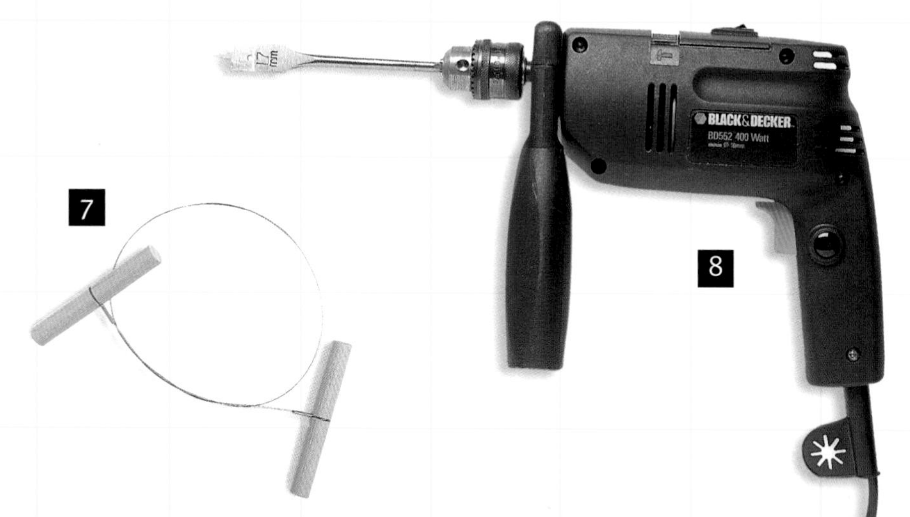

2 Slip trailer; **3** potter's kidneys; **4** potter's needle; **5** side scraper; **6** craft knife; **7** potter's clay-cutting wire; **8** electric drill.

Ceramic tiles are made from clay which has been fired in a kiln, a process which transforms it into a harder, more durable material. The numerous clays available from craft pottery suppliers vary greatly according to the uses for which they are intended.

The Clay Body

Natural clays occur in a wide range of colors and consistencies. For ceramic use, most have other materials added to them, such as fine sand or grog, to modify their texture for various processes.

Some natural clays are usable in their raw state, but most are first refined and have other materials added to them before they are used in ceramics. These reconstituted clays, known as "clay bodies," are generally composed of three elements: a granular material that adds texture and porosity, clay minerals that give the body plasticity, and chemical fluxes that melt and bind the body together.

Clays were historically classified by the items that they were used to make, such as "pipe clays," "brick clays," and "fire clays." Today, they are generally described by their color, the temperature range to which they should be fired, and the making process for which they are designed.

In their natural state, clays range in color from dark gray and red-brown through to light cream or white. These colors modify during firing, with dark colors generally becoming lighter. Black or gray clays in particular show a marked change, becoming tan or light red. Clay bodies can be purchased with added stable ceramic pigments that will maintain their color when fired, but they are expensive and often require extra health and safety measures in their handling. Whatever color clay you use to make your tiles, it can be changed with a coating of colored glaze or slip (see page 60).

The "Maturing Range"

As a clay is fired, its various components gradually fuse together through processes of sintering and melting. Insufficient fusion leaves the fired body weak and too porous. Too much melting causes the resulting ceramic to be highly vitreous and brittle. There is therefore a recommended range of firing temperatures for each clay body, known as the "maturing range," which depends on the chemistry of the body. Two main ranges are recognized: bodies maturing below 2,012°F (1,100°C) are known as earthenware bodies and above 2,192°F (1,200°C) as stoneware bodies. Three types of body that do not readily sit within these categories are Raku, porcelain, and bone china. The Raku body is designed to be low fired and porous with a good thermal shock resistance, whereas porcelain and bone china bodies are fired until they are highly vitreous and totally impervious.

Grog

Within the earthenware and stoneware firing ranges, individual bodies are designed for various processes. A clay suitable for hand-building or sculpture will have a high proportion of granular material such as fine sand or grog (a ground-up ceramic), whereas a body designed for throwing requires a finer smoother texture with a more plastic nature. Grog reduces the overall shrinkage of the body and gives it an open porous texture. With large thickly made objects, this helps the body to dry evenly without warping.

Grog comes in a variety of grades and a variety of sources from pure calcined china clay (molocite) to recycled furnace bricks. The coarsest grades are added to a body to give a visual texture, while finer grogs are used in throwing or press molding bodies (see pages 75 and 89), and in general most bodies will contain a mixture of grades. The main disadvantage of adding grog or sand to a body is that its presence on the finished surface can cause blemishes on the overlying glaze.

Uneven Shrinkage

In tile making, one of the main problems is preventing the tile from warping as the clay shrinks during drying. If the shrinkage is greater on one surface or in one direction, the object will dry out of shape.

Uneven shrinkage is caused by two factors: clay particle alignment and uneven drying. Clay particles, in effect minute platelets, move closer together as the water between them leaves the clay, with the shrinkage greater between their flat surfaces than between their edges. If they are all aligned in the same direction, the overall shrinkage will be greater in one direction than the other. Compression of the clay during the making of a tile causes the particles to align with their flat surfaces perpendicular to the direction of the pressure, whereas applying a shockwave to the clay, by throwing it into a mold or hammering it, tends to randomize the orientation of the particles. Therefore the making process can impart a tendency in the tile for it to warp (known as clay memory), which will be lessened by increasing the proportion of nonshrinking granular material in the body.

Uneven drying occurs if a tile is left to dry on a nonabsorbent surface, when the face exposed to the air will dry more rapidly and shrink faster than the lower surface, causing the tile to curve upward at the edges. As it continues to dry, the hardened upper surface prevents the tile from re-flattening as the lower surface contracts. Grog or sand within the body help to counter uneven drying by providing a more open and porous fabric through which water can migrate from one surface to the other. Using a heavily grogged body (40-50% grog) will help you make a flatter, squarer tile.

Regular turning of the tile during drying can balance out the uneven shrinkage, but this is labor intensive. The answer is to slow down the drying rate of the top of the tile or speed up the drying rate of the bottom face, or to do both.

The drying rate of the top will vary with the ambient humidity and the rate of airflow over the surface, and can be greatly slowed down by covering the tile directly with plastic or placing it within a purpose-made drying box.

To speed up the drying of the lower surface, the tile must be placed on an absorbent surface. This can either be a thick block of dry material such as plaster, which will absorb all the moisture, or a strong permeable material that allows the moisture to pass through it. For the latter you can use a fire-retardant insulating board made from "autoclaved" calcium silicate.

Uneven drying of a clay body can cause warping and shrinkage. One way to avoid this problem is to use a purpose-made drying box, which regulates the rate of airflow over both top and bottom surfaces of a tile.

If you are making a large number of tiles, a wooden drying rack is useful. This can be constructed very simply but must be strong enough to support the weight of the wet tiles without bending or warping.

To check the shrinkage percentage of a clay, mark a line on a freshly made tile and remeasure it after firing. The number of millimeters it has shrunk will represent the percentage.

This comes under several trade names and in various thicknesses with differing qualities of surface finish. With small wall tiles a 3⁄16-inch (4mm) thick board made for infilling soffits will do but with larger tiles a more robust board is needed. To prevent the board from becoming saturated, it must be raised on kiln props or bricks to allow an air-flow underneath, so it must be strong enough to support the weight of the wet tile without warping. If you are producing a large number of tiles, a wooden drying rack can be constructed to hold them in batches.

Shrinkage Tests The total shrinkage of a clay will depend on the top temperature to which the body is fired. To be able to make a particular size of tile, a shrinkage test must be carried out to establish the initial size to cut the plastic clay. The shrinkage percentage can be easily worked out by marking a line 4 inches (100mm) long on a freshly made tile. This line is then re-measured after firing, the percentage simply being the number of millimeters the line has shrunk. Some tile-making methods such as extrusion (see pages 84–8) tend to create a larger shrinkage in one direction than the other, so remember to check for this.

"Wedging" the Clay

If a body with the right texture or color is not available, or if you simply want to even up the consistency of a clay, then it is simple to mix two clays by hand. The process, known as "wedging," superficially resembles kneading bread, but whereas a baker aims to trap air within the dough, a ceramicist endeavors to exclude it, as an air bubble can cause the tile to fracture as it expands during firing.

Wedging should be carried out on a clean absorbent surface such as a block of wood or smooth stone. Over-wet clay can be gradually dried out by this process, but to avoid drying out a clay that is of the right consistency, the work surface should be dampened.

First, thin slabs of the clays to be mixed are slammed together so that any air between them is squeezed out. The far right-hand corner of the clay block is then lifted into the vertical position with the left hand and pushed down into the center of the block with the heel of the right hand, while still held with the left. This is repeated 20 to 30 times so the clay forms a spiraling cone. The process uses a rhythmic, rocking action, with the weight of the upper body

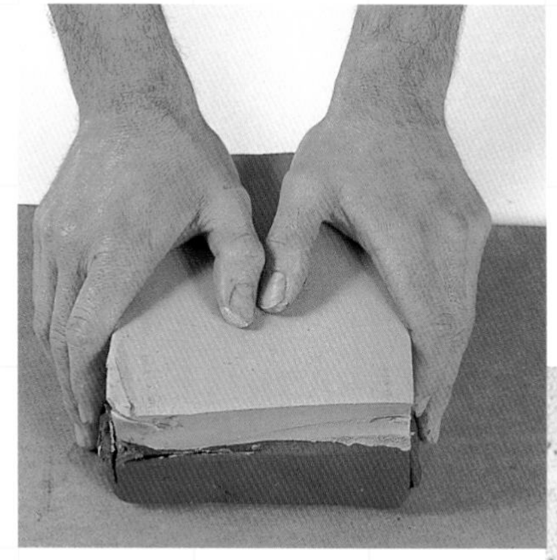
STEP 1 "Wedging": slam together thin slabs of the clays to be mixed.

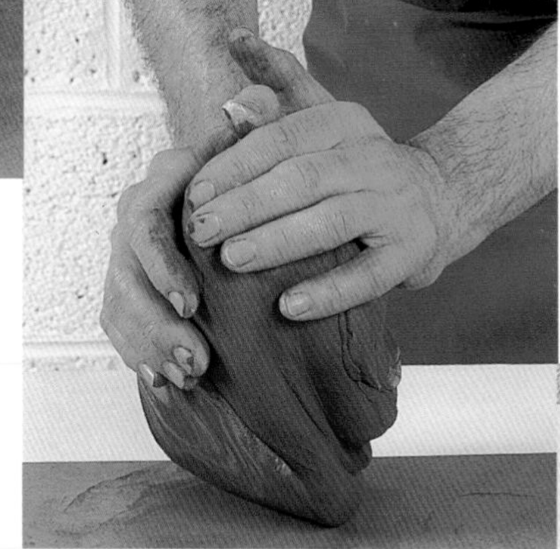
STEP 2 Work the clay into a spiral with a rhythmic kneading action.

pushing the clay down into the block rather than the arm muscles. A low work-bench at about mid-thigh height allows you to keep your arms straight and eases any strain on your back.

To check on the evenness of the mixing, slice the block open with a clay-cutting wire (see page 51) and then continue wedging until the block appears homogenous.

To square off this spiral wedge, or to mix small quantities of clay, a different action known as "ox-head" wedging can be used. Here both hands are used to rotate the block and then to push the clay down and in toward the center. This leaves a raised ridge of clay between your thumbs, and the block quickly takes on the shape of an ox head.

Casting Slips

Most smooth clay bodies are also commercially available as casting slips. These are high-density suspensions of clay particles in water which are used for casting clay forms in plaster molds (see pages 65–6). They are made by slaking down plastic clay or dry powdered clay in water along with a balanced mixture of chemicals that prevent the clay particles from forming into loose aggregations and making the slip too viscous. Fine sand or grog, which has been damped down to reduce its absorbency, can be added to casting slips, but always seek advice from the manufacturer of the slip about any adjustments that may need to be made.

Health and Safety Notes

When dried, clay breaks down into a very fine dust that readily becomes airborne and is visible to the naked eye only when brightly illuminated by sunlight. If this is breathed in, it can become lodged within the lungs, and long-term exposure to high concentrations of dust can cause an acute congestive condition known as silicosis. Where exposure to clay dust is unavoidable, a respirator fitted with cartridges approved for use with ceramic materials must be worn. However, it is far better to avoid the formation of dust by taking several precautionary steps. First, you should prevent plastic clay waste from drying out in your work space. Work surfaces and floors should be regularly washed, and dry clay or slip should be cleaned up with a damp sponge or cloth. Dry brushes or home vacuum cleaners should not be used as they simply create airborne dust. When working with any ceramic materials, always wear a smooth plastic apron which can be easily cleaned, as normal woven fabrics can harbor dust. Coveralls made from a specially designed smooth-surfaced synthetic fiber can be purchased through pottery material suppliers.

The only health hazard in the handling of wet clay is the possible irritation of sensitive skin due to low concentrations of chemicals present in the clay bodies. This tends to be more of a problem with manufactured bodies such as porcelain than with natural clays. With regular exposure to an irritant, the skin can become highly sensitized to a particular chemical which might then cause dermatitis (inflammation of the skin) on even a brief contact. This can be avoided by applying a barrier cream to the skin, but if you do have an allergic reaction, the only answer is to wear rubber gloves. Special care should be taken when using casting slips as these contain alkalis which are mildly caustic to the skin.

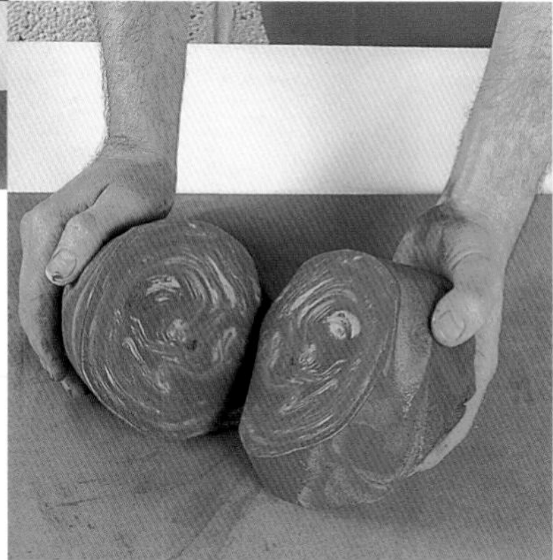
STEP 3 Slice open the block with a wire to check on the evenness of the mixing.

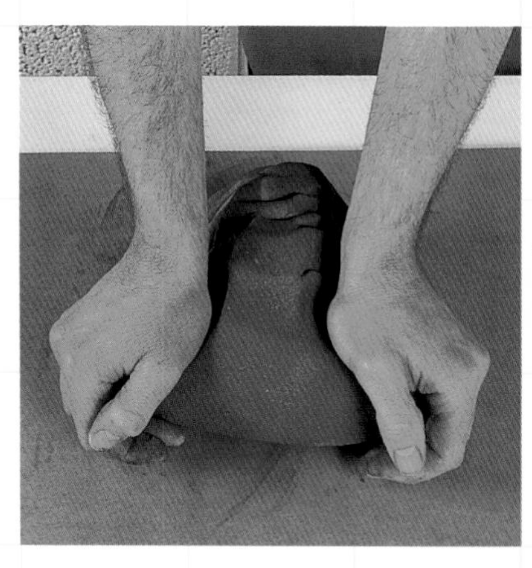
STEP 4 Square off the block, pushing the clay down and in, with the action known as "ox head" wedging.

All clay objects require an initial firing in a kiln to transform them into a hard, durable ceramic, and subsequent firings are used to add layers of decoration, such as glazes and clay slips, on the surface. In essence, a kiln is like any other oven, but the temperatures it reaches are far in excess of any home appliance.

Kilns and Firing

TOP Fuel-burning kilns use either propane or natural gas.
ABOVE Kiln temperatures can be checked with a thermocouple probe connected to a pyrometer.

Most kilns can fire to 2,372°F (1,300°C), and even the lowest melting-point ceramic pigments will not fuse to an object below 1,202°F (650°C). The heat source can be an electric element or the flames of a burning fuel. The main difference between the two is that in an electric kiln, during most of the firing, the atmosphere will have the same balance of gases as the normal air we breathe, with up to 21% free oxygen; but in a fuel-burning kiln, combustion consumes the oxygen, and unless there is an abundant airflow into the kiln, its internal atmosphere will become oxygen-starved. This fundamentally affects the chemical reactions occurring both in the glazes and the clay body, and some pigments will produce a completely different color under these conditions. An oxygen-starved firing is called a "reduction firing," and many decorative glazes will only produce their desired effect in this atmosphere. Reduction firing should only be attempted in a well-ventilated area or with a kiln equipped with an extraction system in the flue, as dangerous gases can be produced.

The ceramic projects in this book are all produced using oxidized firings in an electric kiln. The same finishes can be achieved in a fuel-burning kiln with an adequate through-flow of air.

Kiln Temperatures

The firing of pottery produces chemical and physical changes in the clay body and the glaze which require both high temperatures and time to take place. It is a question of the amount of energy (heat work) transferred by a kiln to its contents rather than simply the maximum temperature reached. To be able to repeat the result of a firing, you need to be able to repeat the same firing cycle. This is described in terms of periods of fixed rate temperature rises called "ramps," and periods when the kiln is held for a certain time at a constant temperature, referred to as "dwells" or "soaks."

Most ceramicists rely on recording the rise and fall of the kiln temperature with a thermocouple probe connected to a pyrometer. To be able to transfer your results to a different kiln or to alter the firing cycle, but still achieve the same result, you have to quantify the actual heat work done to the ceramics. This is tested by placing pyrometric cones (small elongated pyramids of ceramic minerals) in the kiln. These soften and gradually bend over after they have reached a particular temperature at a constant specified heating rate. The same cone will soften at a slightly lower temperature if heated at a slower rate or will stay firm to a higher temperature if heated more rapidly. If cones are overfired, they rapidly slump and form a molten glass. Various manufacturers produce pyrometric cones to cover the entire range of ceramic firing temperatures in intervals of 68–77°F (20–25°C).

Cones should be placed in the kiln at an angle 8 degrees from the vertical, with their bases embedded in a small lump of clay, and positioned as far inside as possible while still being viewable through a spyhole in the kiln door. In each firing, three consecutively rated cones from the range are placed alongside one another. The center one should be the target temperature and those on either side for temperatures 36°F (20°C) above and below that target. During the firing, the interior of the kiln will be incandescent, so care should be taken, both when opening a spyhole and looking through it, to avoid injury from radiant heat or streams of hot gas. Special eye protection in the form of green tinted goggles must be worn to prevent damage from infrared radiation.

Firing Programs

The first firing of a clay body, known as the "biscuit firing," leaves the object hard and easy to handle but still porous enough to be glazed (see page 60). The second firing, known as a "glaze firing," is generally to a higher temperature than the biscuit. In industry the situation is often reversed, with the biscuit firing being to a higher temperature than the glaze firing. The first firing is then known as a "bisc" or "bisque firing" and the second as a "glost firing." The term "bisque" has in recent years become synonymous with "biscuit."

Biscuit Firing Cycle

The main physical and chemical transformations of the clay body occur during the biscuit firing. The firing rate must therefore be slow enough for these reactions to take place and to prevent the physical strain which is put on the clay fabric from breaking it apart.

Before firing, the clay object must be thoroughly dried. Up to 40% by weight of a soft plastic clay is water, about half of which is lost during air-drying. To check that the object is dry enough to fire, either place a small drop of water on its surface or lightly scratch the surface with your fingernail. The water drop should be rapidly absorbed into the clay body and the sub-surface exposed by the scratch should appear light and dry.

At the start of firing, the clay must be heated gradually to at least 248°F (120°C) to drive out any water trapped between the clay particles. The removal of remaining water chemically bonded around and within the crystals of the hydrous clay minerals represents the beginning of the irreversible ceramic change. This begins between 572–752°F (300–400°C), reaches a maximum at around 1,112°F (600°C), and is generally complete by 1,292°F (700°C). From here up, carbon and sulfur burn out of the body, liberating carbon monoxide, carbon dioxide, and oxides of sulfur (see **Gas and Fume Emissions**, below). Both elements need to be removed from the body to avoid problems in subsequent glaze firing, so the biscuit cycle is often slowed toward its end to allow time for this burnout to be completed.

The top temperature chosen for a biscuit firing will be decided by the type of clay, the level of porosity required if a glaze is to be subsequently applied, and such factors as craze resistance. The lowest feasible limit to produce a relatively soft ceramic is 1,382°F (750°C). A practical range for the clays commonly used in craft pottery is 1,742–1,832°F (950–1,000°C). If heating is continued beyond the maturing range of a clay (see page 52), melting or vitrification of the body will cause it to slump. For all projects in this book where a glaze is subsequently applied to the biscuit, a temperature of 1,796°F (980°C) has been chosen. With unglazed projects, a harder or higher biscuit firing of 2,021°F (1,105°C) is used. The firing is always held (soaked) at the top of the cycle to ensure that all the material within the kiln has reached an even temperature.

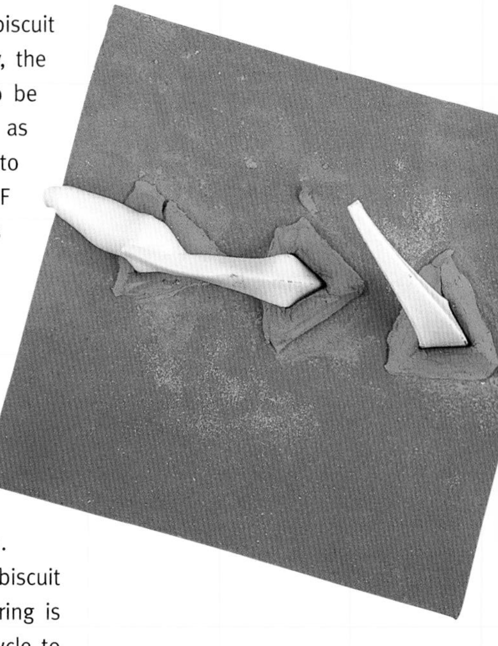

Generally the cooling is uncontrolled, but care must be taken that a ceramic is not cooled too rapidly through 1,063°F and 439°F (573° and 226°C). Physical changes occurring in the clay at these temperatures cause a sudden expansion on heating and a equally sudden contraction on cooling, which can result in a rigid fired body cracking apart ("dunting"). With any firing you should leave the door closed until the

TOP Pyrometric cones are placed in a kiln to test the amount of heat work transferred to ceramics during firing.
ABOVE After the cones reach certain temperatures at a specified heating rate, they will soften and bend over.

Kilns are available in many shapes and sizes, so it is advisable to gain experience of using one before considering a purchase. A small top-loading kiln, as shown here, would be suitable for a home ceramicist.

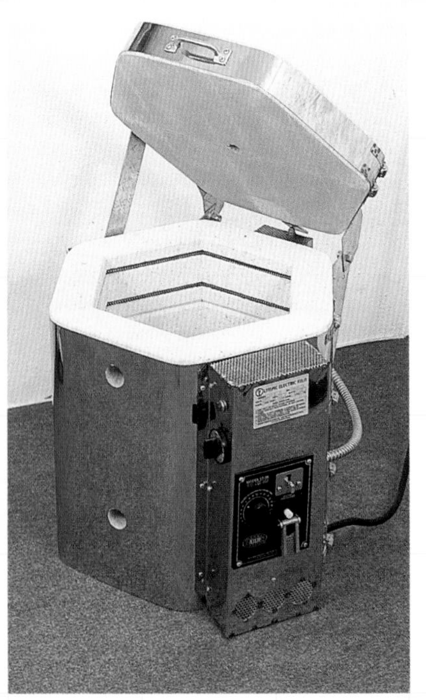

temperature has dropped to below 212°F (100°C). Even at this temperature, the ceramics and kiln furniture (see below) will hold onto their heat for some time, and protective gloves should be worn before handling them. The firing cycles used for the low and high biscuit firings in this book are as follows.

Low Biscuit Firing This program is suitable for ceramics which will later be glazed.

- Room temperature to 248°F (120°C) at 54°F (30°C) per hour.
- 248–932°F (120–500°C) at 90°F (50°C) per hour.
- 932–1,796°F (500–980°C) at 180°F (100°C) per hour.
- Soak the kiln for 30 minutes at 1,382°F (750°C) and for 30 minutes at the top temperature. These temperature ramps are maximums, and if your kiln heats at a slower rate, it will not effect the outcome.

High Biscuit Firing Use the same firing cycle as for low biscuit, but continue until the top temperature reaches 2,021°F (1,105°C).

Glaze Firing

Glazes are applied to the biscuit-ware as an insoluble powder suspended in water (see page 60). The water is absorbed into the ceramic, leaving the powdered glaze on the surface. During the glaze firing, this absorbed water must first be evaporated off without dislodging the layer of powder adhering to the surface. From room temperature to 212°F (100°C), the kiln should therefore be fired at the same rate as a biscuit firing. If both the glaze and biscuit are fired with normal oxygen-rich atmospheres, there will be no further chemical change within the fabric of the body until the maximum temperature of the biscuit firing is exceeded. However, the physical changes in the clay will recur every time the ceramic is heated or cooled through 439°F and 1,063°F (226° and 573°C), creating high stresses in the body. The rate of the glaze firing must therefore remain slow until these inversion temperatures are passed. Any wax resist applied to the biscuit ware will also burn off during this period of the firing, producing noxious fumes (see **Gas and Fume Emissions** below).

The changes that occur in the glaze layer will depend on the type of glaze being used. Oxidized glazes can usually be fired rapidly from 1,112°F (600°C) up to close to their maturation temperature, when the firing is slowed down to allow all the materials to melt completely. A stoneware glaze is often fired at a slower rate than earthenware once the biscuit temperature has been exceeded to allow for interactions between the glaze and body which increase as the temperature rises above 2,012°F (1,100°C).

Earthenware Glaze Firing This is used as a second stage firing in a number of projects.

- Room temperature to 212°F (100°C) at 54°F (30°C) per hour.
- 212–1,112°F (100–600°C) at 135°F (75°C) per hour.
- 1,112–1,832°F (600–1,000°C) at 270°F (150°C) per hour.
- 1,832–1,976°F (1,000–1,080°C) at 90°F (50°C) per hour or, alternatively, soak the kiln at 1,976°F (1,080°C) for 30 minutes.

Stoneware Glaze Firing The Wax-resist Panel Design project (page 101) uses this program.

- Room temperature to 212°F (100°C) at 54°F (30°C) per hour.
- 212–1,112°F (100–600°C) at 135°F (75°C) per hour.
- 1,112–1,832°F (600–1,000°C) at 270°F (150°C) per hour.
- At 1,832°F (1,000°C) soak for 1 hour. (This is necessary with some stoneware clays as they contain a large amount of carbonaceous material that might not have fully burned out during the biscuit firing.)
- 1,832–2,192°F (1,000–1,200°C) at 180°F (100°C) per hour.
- 2,192–2,300°F (1,200–1,260°C) at 90°F (50°C) per hour, or soak the kiln for half an hour at 2,300°F (1,260°C).

Gas and Fume Emissions from Kilns

As described above, at various stages during a firing, gases and fumes are produced within a kiln which must be allowed to escape through a vent or flue. In high concentration these gases can be both noxious and poisonous. No kiln of any type should be fired in a room where people are working, and the room must be well-ventilated or equipped with an extraction system. Fuel-burning kilns produce waste gases throughout the firing and will either have a chimney fitted directly onto the vent from the kiln or a hood directly above the flue leading into a chimney or an extraction system. Closing off the flue reduces the amount of oxygen entering the kiln. With an electric kiln, the vent can be partly closed off to conserve heat above

932°F (500°C), when all the water has been driven out of the kiln but small volumes of fumes are still being generated by the burning off of carbon and sulfur.

Choosing a Kiln

Kilns are available in many shapes and sizes, and they can be expensive to buy and install. It is therefore advisable to gain some first-hand experience of working with a kiln at a ceramics class before attempting to purchase one. Alternatively, your local pottery or pottery materials supplier may be able to offer a firing service. Always check what temperatures they fire to before starting a project so you can choose the right body and glazes.

When buying a kiln, you should consider several factors including size, ease of use, the power and fuel supply required, and the cost of installation and firing.

To use the space within the kiln efficiently, you need a system of movable shelves ("kiln batts") and supports made from a refractory (high melting point) material. Most kilns will come with a set of custom-made furniture, but shelves can be bought separately in a large variety of weights, sizes, and shapes, and suppliers will generally cut them to order.

Shelf props, used three to each shelf, are hollow tubes with either flat or castellated ends. They are available in a variety of heights, and the castellated props can be interlocked with each other to increase the range.

Placing Ware in a Kiln

When packing a biscuit kiln, objects can be placed in contact with each other and directly on the shelves. Tiles fired lying flat should be bedded on fine white sand (silica sand) to reduce the friction between the shelf and the clay body as they shrink. However, where possible, tiles should be biscuit fired standing on their edges. If they lie flat, the surface in contact with the shelf tends to bake harder and will contract slightly more than the upper surface, causing the tiles to bow upward.

In a glaze firing the glazed surfaces must be kept apart and not allowed to come into contact with any of the kiln furniture. A molten glaze will stick to any surface it touches, and if applied thickly, it can flow down over an unglazed edge. Kiln furniture can be protected from glaze contamination by applying a thin layer of high alumina wash (batt wash) to the surface.

With earthenware pottery the undersurface of the object is often glazed to seal in the porous body. It is therefore necessary to support the ware away from the kiln shelf by the use of stilts or spurs. These small ceramic stands hold the object on narrow spikes which become embedded in the glaze but can easily be broken away after firing is completed.

Flat glazed objects must also be held away from the shelf to ensure a more uniform heating. For stoneware firings the undersurface must be cleaned of glaze and then the object can be placed upon blocks (bits) or purpose-made stands ("tables" or "chairs").

An efficient way to stack tiles in a kiln is to use cranks. These are frames made of refractory material with three uprights held in position with a base and top plate. The uprights have evenly spaced pegs or lugs which in a low-temperature firing can support the tiles directly. With higher-temperature firings where the tiles might possibly warp, they must be loaded into the crank on small kiln batts.

Movable shelves, known as "kiln batts", are needed within a kiln to carry the ware for firing. These shelves can be built up in stacks with interlocking props for support. The thicker the shelves, the more heat will be required to raise their temperature.

A glaze is a layer of glass that is applied to the surface of a ceramic object and then fused on to it by firing in a kiln. Unfired clay can be decorated with a "slip," which is basically a mixture of fine clay and water. Both glazes and slips are adaptable materials which can be used with a variety of techniques to produce bold decorative effects or a layer of strong color.

Glazes and Slips

Glazes are generally made as insoluble powders that are applied to the ceramic as a suspension in water (a "slop"). Some are available ready-made from pottery suppliers, either already mixed with water for immediate use or as a dry powder, but it is possible to make your own by combining basic ingredients in much the same way as in cooking. Some glaze recipes are given at the back of this book (see page 141), and many others can be found in the wide selection of pottery books currently available. With a basic understanding of how glazes work, you can modify such recipes to suit your purpose.

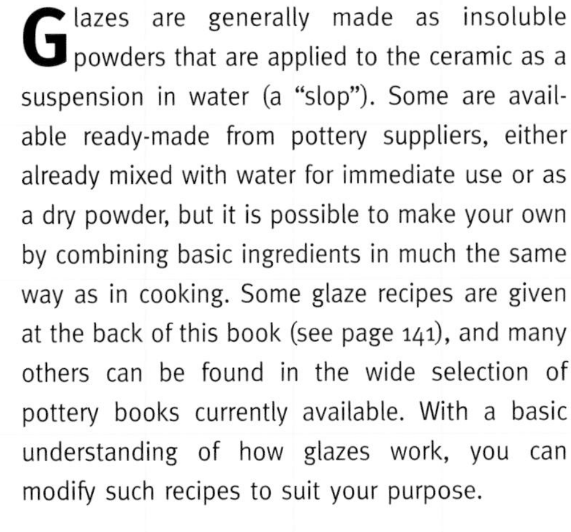

Fired glazes add color and richness to a ceramic surface. They can be purchased ready-made, but you can also mix your own to achieve a particular finish.

Glaze Constituents

The major constituent of all pottery glazes is silicon oxide (silica) or a combination of this with boron oxide (boric oxide). These are known as the "glass formers," as it is their presence in a melt that encourages the formation of a glass, rather than a crystalline solid, on rapid cooling.

Silicon oxide by itself would not melt until well above the normal firing range of most pottery clays, but when other chemicals are added to it, its melting point is reduced. Such chemicals are referred to as "fluxes" and affect the relative hardness of the glass they form. At the comparatively low temperatures used for firing earthenware glazes, the oxides of sodium, potassium, and lead (the "soft" fluxes) are effective. At high stoneware temperatures, when some start to boil out, these must be gradually replaced by the "hard" fluxes of calcium, barium, magnesium, and zinc oxide.

The other important component of most glazes is aluminum oxide (alumina), which stiffens the molten fluid, increasing its adhesion to a ceramic surface; without it a glaze would easily run off an object within the kiln. Alumina and silica are generally introduced into the glaze in the form of clay, which helps to keep the other glaze materials from rapidly settling out.

The appearance of a glaze can be altered with the addition of other chemicals. For instance, opacity can be produced by oxides of tin, zinc, and titanium, and a silky matte finish can be produced by the addition of calcium, magnesium, barium, aluminum, and titanium. (A matte surface with a dry appearance can also be formed from underfiring a glaze, but the surface will be rough and unsuitable for household use.)

Glazes made from raw materials will undergo various chemical changes during melting, and a small amount of time is required after melting for the glaze to settle (mature) to a smooth surface. Most glazes rapidly thicken on cooling and will remain soft and liable to change chemically for at least 720°F (400°C) below their maturing temperature.

Color in Glazes The vast array of colors seen on ceramic surfaces (excluding metal lusters) are produced from just ten elements in various combinations with one another and with other glaze constituents. These are listed below with their most characteristic colors:

- **Iron** amber to brown in oxidation, green to dark brown in reduction firing
- **Copper** turquoise to green in oxidation, deep red in reduction firing
- **Cobalt** blue
- **Manganese** purple to brown
- **Nickel** gray or brown
- **Antimony** yellow in association with lead
- **Vanadium** yellow
- **Chromium** green or pink in combination with tin
- **Cadmium** orange and yellow; in combination with selenium can produce bright reds in reduction firing
- **Depleted uranium** yellow and green

The first eight elements are readily available as oxides or carbonates for direct addition to a glaze or for using as painting pigments. The color of an unfired glaze or pigment will frequently alter in the kiln. To increase the predictability of colors, the ceramic industry has produced a range of stains which remain stable. These are produced in various grades for use either within slips and glazes or as pigments for painting beneath or on top of a glaze.

Frits The components used in a glaze slop need to be insoluble in water, both to decrease the health risk to the ceramicist (see **Health and Safety** notes, below) and to avoid their being drawn out and concentrated on the surface of the glaze layer as it dries. Because the fluxing oxides are water-soluble, they are converted for ceramic use into a fine powder containing a glass former. These insoluble powders are known as "frits," and a variety of compositions are commercially available. It is advisable to avoid using a lead frit in glazes for household pottery, such as kitchen counter tiles, as the risk of lead poisoning is not unknown.

Frits are expensive due to their manufacturing costs, and where possible, naturally occurring minerals that already contain the fluxes should be used. However, most natural minerals do not normally contain a high enough proportion of soft fluxes to be useful in earthenware glazes. Advice on the range of materials available can be provided by your local supplier or through the technical information department of the manufacturers.

Enamels

Enamels are low-melting-point glazes that contain a stable colored stain. They melt and fuse onto the surface of a base glaze at around 1,292–1,472°F (700–800°C). This is above the softening range of most glazes, but with the exception of low temperature reduction glazes (i.e. Raku copper glazes), the base glaze will not be altered by this refiring.

Glaze Preparation

The first stage in making a glaze from a recipe is to place enough water for your chosen mixture in a bucket. As a rough guide, 2 pints per 2 pounds (1.25 liters per kilo) of dry material will suffice. The weighing out of dry materials (see **Health and Safety** notes, below) should be carried out with care and accuracy on a sensitive set of scales or a balance. Recipes are usually given in percentages by weight and can be easily converted to actual weights by assuming each percentage is equal to one ounce or gram. As each material is weighed out, it should be immediately added to the water and checked off the recipe list. Always add the clay content of your recipe to the water first and stir it well to help suspend the other ingredients, which should be added as gently as possible to stop them from sinking to the bottom, where they can form a solid mass.

When the recipe is complete, make sure the water covers all the ingredients and leave the mix to stand for at least half an hour before stirring it to break up any large lumps of powder. The mix should next be stirred through a 60-mesh strainer with a coarse brush (lawn brush) or a rubber kidney (see page 50) and then passed twice through a 120-mesh strainer to make sure it is reduced to the finest particles possible. Extra water will generally have to be added at this stage, and if one of the ingredients proves too coarse for this strainer, a 100- or 80-mesh should be substituted. The usual method is to support the strainer on two strips of wood over a bucket while the glaze is pushed through. To avoid contamination, different strainers and brushes should be used for white and colored glazes.

The consistency of slop to aim for depends on the method of application (see below). It does not matter if too much water is initially added to the mix, as if it is left overnight to

Glazes generally come as dry materials to be mixed with water. Each powdered ingredient should be weighed carefully on a set of scales before being added immediately to the mixture.

Once the glaze mixture is complete it must be stirred through a strainer to make it as fine as possible. Support the strainer over the bucket on two strips of wood and push the glaze through with a coarse brush

settle, the excess can be skimmed off. Once you have established the correct slop density for your application, it can be quantified by measuring the specific gravity with a hydrometer or by keeping a record of the weight of a set volume (i.e. a bottle full) of the glaze.

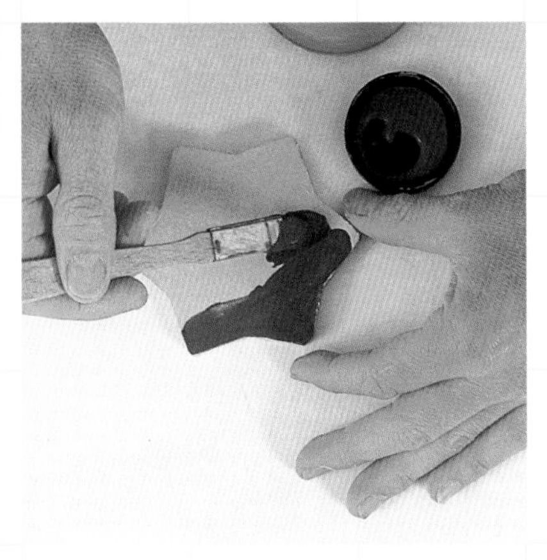

Ready-mixed brush-on glazes are convenient for use if you are decorating only a small number of tiles.

Glaze Application

Glazes can be applied by painting, spraying, or dipping.

Painting Brush-on glazes are thickened especially for application with a brush, and a large variety are available ready-mixed from pottery suppliers. For small-scale use they are convenient and allow you to place several different glazes on the same piece. However, for a large run of tiles, they can be expensive.

Spraying The consistency of a glaze for spraying must be thin enough to allow it to pass through a spray-gun smoothly without clogging. This method must be used in a spray booth, but is only necessary for particularly large pieces or for applying a glaze over unfixed pigments.

Dipping The most convenient technique for applying a glaze to tiles is dipping. This relies on the ability of a biscuit-fired clay body to absorb water. The tile is immersed in a bowl of glaze in a single smooth motion and held submerged while water enters the body, leaving a layer of glaze adhering to the surface. When the tile is removed, it must be held with the finished surface uppermost while any excess glaze drains off.

The glaze can be used in a variety of consistencies with a slightly varying technique. With a thick glaze a substantial deposit can be achieved simply by skimming the finished face of the tile through the surface of the glaze. This will leave heavy wet runs, so the tile must be shaken to even these out. This technique suits a finish where slight unevenness will not be evident. For a fine smooth glaze, a thinner slop should be used so that after total immersion of the tile any excess runs off easily to one edge. If a slightly thicker deposit collects on that edge, the tile should almost immediately be redipped in the opposite direction. The longer the tile is immersed, the thicker the glaze deposit, though the rate of deposition slows down rapidly as the subsurface becomes saturated with water. A thicker deposit can also often be achieved by removing the tile after a brief dip and allowing a few seconds for the damp shine to fade from the surface before redipping it. If the tile is left to dry before redipping, the second layer of glaze will form many small pinholes which must be rubbed down before the tile is fired. The thickness of a glaze deposit can be easily checked by gently pressing your thumbnail or a pin into the surface until it reaches the clay. Most glazes should be applied no thicker than $^1/_{32}$ inch (1mm).

When dipping a thick handmade tile, it is relatively easy to hold the edges between the fingertips, even while wearing rubber gloves, and still avoid leaving unsightly fingermarks. With a thin dust-pressed biscuit tile, this is nearly impossible, and a small wire dipping frame is needed (see page 51). The tile is held on the frame by the very tip of your thumb which should leave only a small mark on the tile edge.

If you are using a thick glaze for dipping, simply skim the finished face of the tile through the surface of the glaze and then shake the tile to even out any runs.

When dipping a large, thick tile it is easy to hold the edges between your fingertips and still avoid leaving marks in the glaze.

A small wire dipping frame is needed when you are glazing thin tiles. Just the very tip of your thumb can be used for support.

Fettling and Waxing

Tiles will stick better to adhesives and grouts if their backs and edges are unglazed. Unwanted glaze can be laboriously scraped off (fettled) after dipping, but a more convenient (and economical) method is to prevent glaze being deposited on these surfaces by coating them with a water-based resist of wax emulsion. When a piece of ceramic is dipped into it, the water enters the clay body, leaving a deposit of wax which, when dry, prevents further water entering the body and therefore stops glaze from being deposited on the surface.

Most wax emulsions are supplied by ceramic material retailers in a thick consistency for painting on resist decoration (see page 101) but can be watered down. The wax should be kept as thin as possible, not only to give a more even coating but to keep the production of noxious fumes to a minimum as the resist burns off during firing (see page 58). With large tiles the resist is most easily applied with a brush or synthetic sponge which should be washed in warm soapy water immediately after use. Smaller tiles can be lowered into a shallow bath of the resist using another simple wire frame (see page 51). Particular care should be taken not to get wax onto the front face of the tile as it is virtually impossible to remove without burning off in a repeat biscuit firing. Very thin tiles can be waxed by being pressed onto a sponge soaked in the resist.

The resist should not be taken more than halfway up the edge of the tile, so after glaze dipping, the edges will still need some fettling. This is done by running the blade of a penknife or a modeling tool around the edge of the tile, working quickly while the glaze is still damp to avoid creating harmful dust (see **Health and Safety** notes, below). The fettled-off waste should be discarded before it dries out, but it should not be returned to your glaze bucket as it is likely to contain scrapings of wax. Any small amount of glaze that does stick to the wax resist can be easily wiped off with a damp sponge before the tile is placed in the kiln.

Decorating Slips and Vitreous Slips

A decorating slip is a fine, smooth suspension of clay in water which can be used on a clay surface. It is applied to the unfired clay, and the shrinkage of the slip and the body underneath must be matched up if you are to keep the slip from cracking and falling off the base tile during drying or firing. In its simplest form, decorating slip is made by adding dry clay to water, but other ingredients including fluxes and colorants can also be added. Most clays contain their own fluxes which hold them together and onto an object when they are fired. A vitreous slip is any clay slip that forms a shiny surface through partial melting. The addition of fluxes to any base clay can produce a vitreous slip, and recipes can be devised to give a complete range from a dry matte decorating slip through to a clay-rich glaze.

Coloring oxides or stains can be used in slips, but slightly higher percentages than are normal in glazes are required to produce a strong effect. Vitreous slips are generally used as the finished surface whereas non-vitrifying slips are subsequently covered with a transparent glaze.

BELOW To prevent unwanted glaze from adhering to the back of a tile, sponge on a thin coating of wax emulstion.

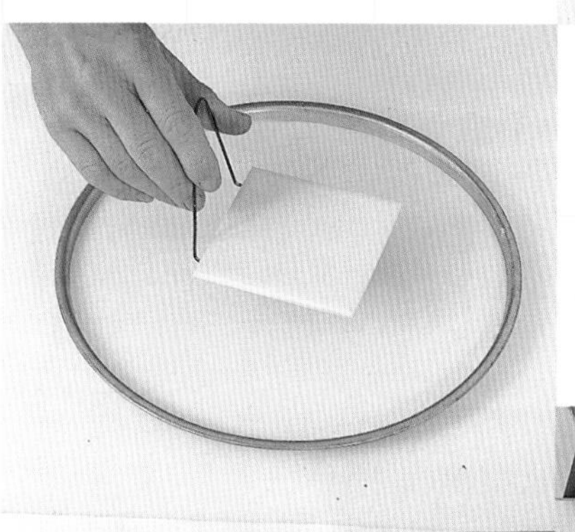

LEFT Small tiles can be lowered into a shallow bowl of wax resist, using a simple wire frame made from a coat hanger.

RIGHT Remove excess glaze from the edges of a tile by running a penknife or similar blade around while the glaze is still wet.

LEFT Before placing your tile in a kiln, give it a wipe with a damp sponge to remove any final traces of glaze still sticking to the wax resist.

Decorating slip can be applied to small tiles in the same way as a glaze. Skim the tile through the surface and shake it to settle out any runs.

Applying a Slip When you apply a slip to an unfired object, care must be taken to avoid saturating the underlying clay as this can cause it to break up. The thickness of the slip is critical and must be adjusted to suit the dryness of the tile. With small tiles the slip is applied in a similar way to a viscous glaze by skimming the tile through the surface and then shaking it to settle out any runs on the surface. Small blemishes will disappear almost completely as the slip dries. Larger tiles are coated by pouring a consistent stream of slip over the surface while being supported over a bowl on a simple stand made from a triangle of board with two small pegs. The edges of the freshly slipped tile are easily cleaned up with a small natural sponge.

Health and Safety with Glaze Materials

Nearly all the materials used in the making of glazes, slips, and pigments can be harmful by inhalation, ingestion, or absorption through the skin. Most are likely to cause only mild irritation or illness, but in a few cases misuse can have much more serious effects. This does not mean that ceramic work cannot be carried out safely if you handle materials correctly. All the necessary precautions you need to take are either common sense or basic good housekeeping. Information on the level of toxicity of each material and what is an acceptable exposure to it can be obtained from your ceramic supplier. However, small children and pregnant or nursing mothers are particularly vulnerable and should take care to avoid any exposure to toxins or solvents.

For coating larger tiles with slip, support them over a bowl on a simple stand and pour the slip from a pitcher in a steady stream.

Clean the edges of a freshly slipped tile by wiping them with a small sponge.

The inhalation of fine airborne dust should be avoided by wearing a respirator approved for ceramic materials. In general, work with dry powders should be kept to a minimum, and they should be added to water as soon as they are weighed. Dry mixing must always be avoided, and any spillages should be immediately cleaned up with a damp sponge. The spray application of a glaze or pigment will produce both minute airborne droplets and a fine deposit of dust on surrounding surfaces. To avoid contamination of the work space, spraying must be carried out in an appropriate spray booth with an exhaust system. This equipment is expensive, but is available at most educational establishments.

A standard rule is never to eat, drink, or smoke in the pottery. It should become second nature to wash your hands thoroughly after working in the studio, and the use of disposable paper towels will avoid recontamination.

The risk of absorption through the pores of the skin is greatest with very fine or soluble materials when carried by water. The metallic compounds that produce color in glazes and pigments are often extremely fine and toxic. If contact with the skin is likely when handling wet materials, rubber gloves must be worn. Fine surgical gloves are available from a pharmacy, but care should be taken when putting them on as they are easily split. Barrier cream can be applied as a secondary defense, but this is not enough in itself. Cuts or abrasions are particularly vulnerable and should be covered with waterproof bandages.

All ceramic materials, whatever their toxicity, should be stored safely in clearly labeled airtight containers, out of reach of children and away from any food preparation area. A container should be immediately re-sealed whenever material is removed, and the outside surface should be cleaned off with a damp sponge.

The simplest way to form a tile, whether it has a flat surface or a sculptural relief, is to model it out of plastic clay. However, to repeat a complex design or to try to make perfectly even surfaces can be laborious. It is preferable to make a reusable mold from a prototype model.

Making and Using Plaster Molds

Weigh the plaster powder into a dry bucket and fill a second bucket with the corresponding volume of water.

Molds are used either for pressing plastic clay into, or for casting from a liquid clay (slip). Both press molds and slip molds must be able to absorb moisture from the clay, and the most versatile material from which to make them is plaster. This comes as a powder, and when it is mixed with water a chemical reaction binds the particles together. When set, the plaster will occupy only a minutely larger volume than when it was first mixed and therefore a plaster mold accurately reproduces the original model.

Different types of plaster are manufactured for various uses, but the type of most use to the hobby tile-maker is "potter's plaster," which is readily available from craft materials suppliers. This has a long enough setting time to allow it to be mixed by hand and poured over a model, and although it is not as fine or as durable as some other plasters, it is strong enough to make molds for short-term use. The lifetime of a mold will be dependent on the care with which it is used and the fineness of the clay being cast; coarse clays will gradually abrade the plaster.

A manufacturer will recommend what weight of plaster should be mixed with what volume of water. It is always advisable to do a small test mix first. If you make the mixture too rich, the plaster will thicken and set before you have time to pour it properly. If you make it too thin, the mixture will take too long to set and the final mold will be weak.

When mixing plaster, weigh the appropriate quantity of powder in a dry bucket and put the corresponding volume of water into a second bucket. Gradually add the powder to the water by gently sprinkling it over the surface. Much of the volume of dry plaster is trapped air, and when it is fully mixed with water, it will only increase the initial volume of the water by approximately 50 percent. As the powder is added, it will mound up, forming small islands which gradually sink into the water as the air-bubbles escape and the powder becomes wet. This process, known as "slaking," takes three or four minutes, although it can be hastened by tapping the outside of the bucket. When the mix is fully slaked, the wet plaster will fill the bucket to within an inch (2.5cm) of the water's surface.

Gradually add the powder to the water by gently sprinkling it over the surface.

Wearing a rubber glove, stir the plaster mixture with your hand until it has thickened to the consistency of double cream.

Wait until bubbles are no longer rising rapidly to the surface, then stir the mix continuously until it is ready to pour. This is indicated by a slight thickening from a light to a heavy-cream consistency. The most effective method is to wear a rubber glove and stir with your hand below the surface to avoid introducing any more air bubbles. The mixture will warm slightly as it starts to react. Do not pour too soon as this will allow some of the water to separate out before it sets. Once the setting begins, the mix will thicken rapidly, and if it is poured too late, it will be too viscous to allow trapped air bubbles to escape from the surface of the model.

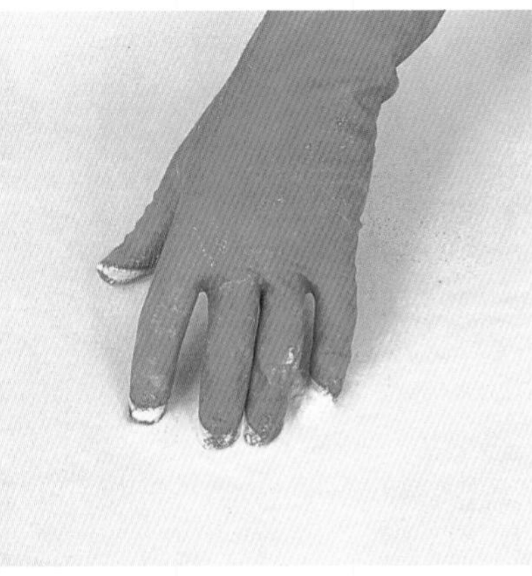

When the plaster has been poured, gently move your hand through the surface to release air bubbles.

If your prototype is made of a delicate material such as soft clay, then avoid pouring the bulk of the plaster directly onto it. To help release trapped air bubbles from the model's surface, the plaster should be gently agitated before it sets. This can be done by moving your hand in the surface of the plaster or indirectly by gently vibrating the workbench.

If more than one pour is required to complete a mold, try to make sure the model has a thin, even coat over its entire surface as the junctions between pours will leave a mark on any subsequent clay casts. For good adhesion between each pour, leave a roughened surface at the top of the previous pour to act as a key for the next. With large molds, extra strength can be added by putting strips of open-weave scrim across the boundaries between pours. If the table you are working on is horizontal, then the final pour should find its own level. The resulting mold must be an even thickness of 1½ inches (40mm) on all sides of the model to ensure even drying of the casts. The mold will be ready for use once it has fully dried, and the time this takes will depend on the ambient temperature and humidity. Warming the mold will speed up the drying process, but if its temperature rises above 104°F (40°C), the plaster will start to revert to the dry powder from which it was made.

If you are making a two-piece mold, the model must have no undercuts. The shading on the diagram shows how indents must be filled in with clay.

If you have made more plaster than you need, it must not be washed down a sink. Either transfer it into a malleable container and swirl it around to form a thin layer that can be easily cracked out when it has set, or wash your bucket out in a large water-filled bowl. In excess water the plaster will form a weakly consolidated deposit on the bottom of the bowl that can be scraped out when water is drained off.

Model Preparation

Models can be made out of almost any material as long as it is not too compressible, too porous, or too highly textured. You can use a manufactured object such as a stone tile or natural objects such as a piece of fruit or a fossil. If you want to sculpt your own model, you can use clay, plaster, or dense styrofoam. Whatever the material, you must ensure two things: first, if you are making a simple press mold or a two-piece slip mold, any relief must have no undercuts. With natural objects, heavy texture that might cause small undercuts can be reduced by applying clay to the surface. Second, any porosity must be reduced by soaking the material in oil or water and then treating the surface with three coats of soft soap, each coat being applied thickly and then wiped off again to leave a thin film.

Clay models can either be pulled out while they are still soft or left to dry within the mold until they contract and fall out. Models made of other materials might remain held in the mold by suction. These can be released by holding a heavy, broad object against the back of the mold and hitting it with a hammer.

If any plaster chips off the casting surface of the mold, or you find any bubble holes, repairs can be made with a small mixture of plaster followed by a rubdown with fine emery paper. As the first cast from a mold is likely to be contaminated with fine plaster dust, it is best to consider it as expendable and pull it out of the mold as soon as possible.

Making and Decorating Tiles

Handmade tiles have an individuality and vitality that cannot be matched by any commercial product. The projects that follow demonstrate various techniques for making your own tiles – by working with plastic clay or casting liquid clay (slip) in a mold, and also using paper-clay, a new material that is rapidly gaining popularity with ceramicists. There are exciting ways of decorating your tiles, using both traditional and modern methods. Some of the projects need only the simplest of tools, such as a wooden board and a rolling pin; and while others require specialized pottery equipment, this is readily available in school or college workshops. All the tiles described must be fired in a kiln.

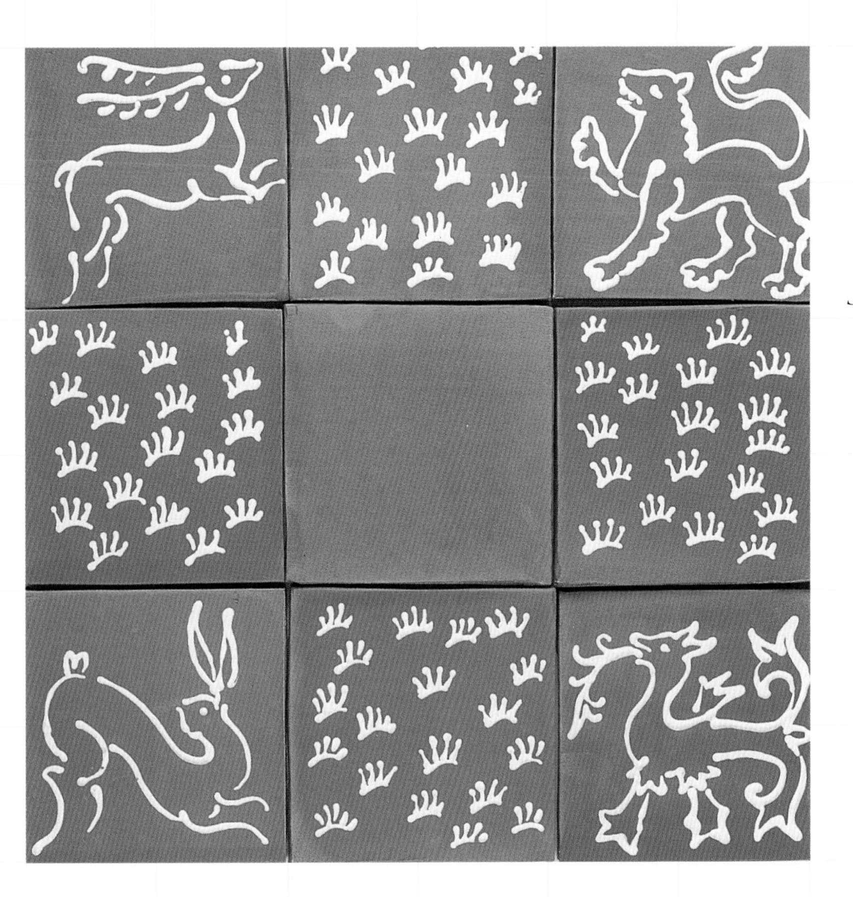

Slab cutting is a simple method for making small thin tiles with the minimum of equipment. Such quickly produced tiles make an ideal base for practicing slip trailing, one of the most fluid of decorative techniques, which is seen in most European cultures in the decoration of rustic pottery.

Slabbed Tiles

Equipment & Materials

(TO MAKE APPROXIMATELY 12 5 × 5-INCH (127 × 127MM) TILES)

- *10 pounds (4-5kg) red terracotta clay (20% grog)*
- *Absorbent wooden board*
- *2 ½-inch (12mm) thick strips of wood*
- *Potter's clay-cutting harp*
- *Sheet of cardboard or thin plywood*
- *Sharp knife*
- *Decorating slips in 2 contrasting colors*
- *Wide bowl*
- *Slip trailer*

Slip trailing involves piping one color of liquid clay onto the wet surface of another with a trailer consisting of a collapsible bag and a narrow nozzle. Although this technique takes time to master, once you have control of it, you will be able to produce decoration with more vigor and life than is possible with virtually any other method. Find a design to copy and try trailing on paper first, to get the feel of it.

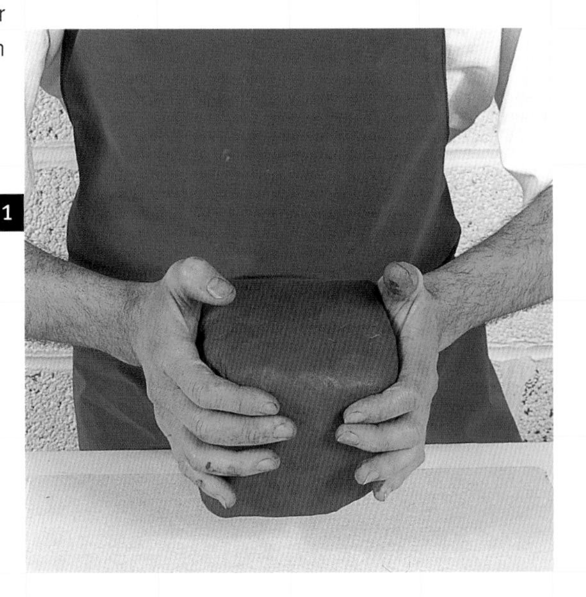

1 Knock the clay into a block with a roughly square profile by dropping it repeatedly onto a dry absorbent board, alternating the face on which it is dropped. Once it is squared, check that the cross-section is large enough for the size of tile you wish to cut. Broaden it if necessary by dropping it on each end before gently re-squaring it.

3

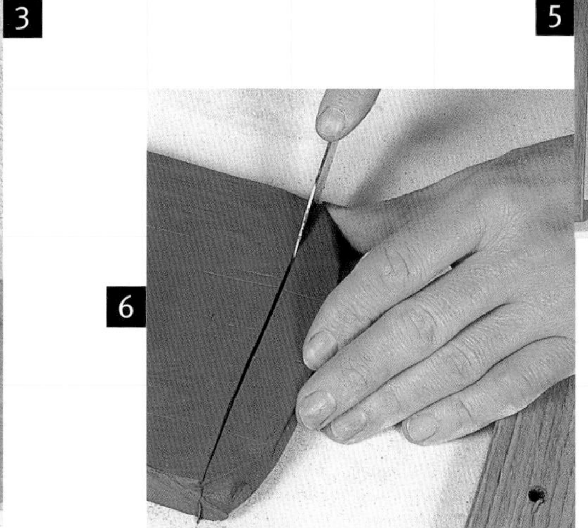

6

5

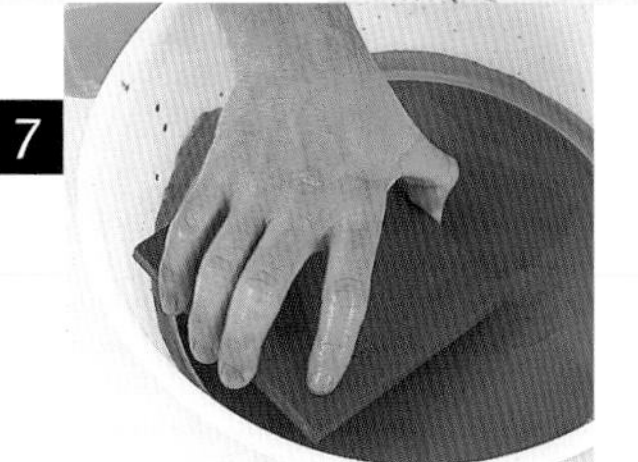

7

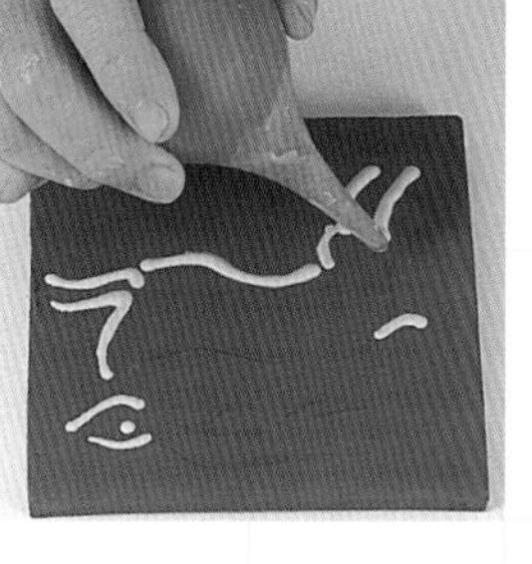

8

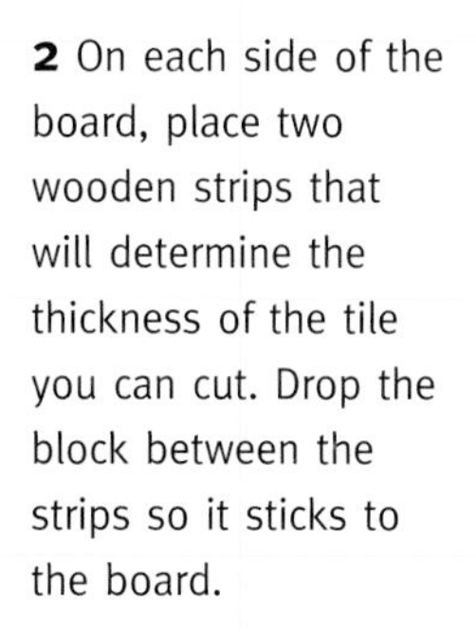

2 On each side of the board, place two wooden strips that will determine the thickness of the tile you can cut. Drop the block between the strips so it sticks to the board.

3 Take the clay-cutting harp and run it along the two strips of wood, cutting a tile from the bottom of the block.

4 Peel off the block, leaving the tile behind stuck to the board. If the tile will not separate, either dust the cutting wire with talc and re-cut or gently tease the tile away as you lift up the block. (If it still will not separate, either use a slightly firmer bag of the same clay or try a more grogged body.)

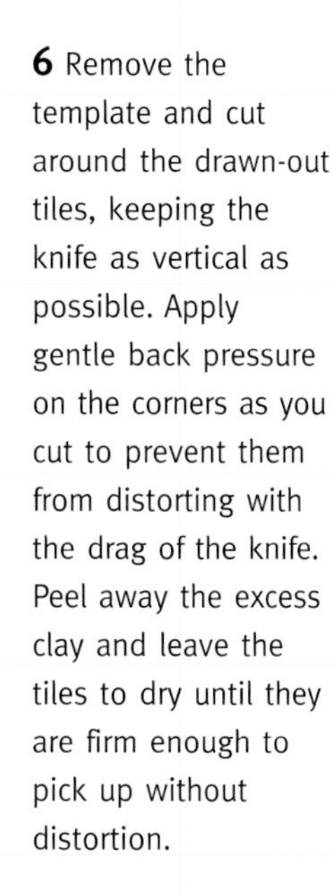

Move the block to another part of the board and continue cutting sections until the board is full.

5 Cut out a template from cardboard or thin plywood to the size required for your tiles, allowing for shrinkage during firing (see page 54). Mark a tile on every cut section by placing your template on the clay and drawing lightly around it with the tip of a sharp knife. If you make a plywood template, it is useful to mount a wooden handle on the back so you can pick it up without marking the slab beneath.

6 Remove the template and cut around the drawn-out tiles, keeping the knife as vertical as possible. Apply gentle back pressure on the corners as you cut to prevent them from distorting with the drag of the knife. Peel away the excess clay and leave the tiles to dry until they are firm enough to pick up without distortion.

7 Pour the slip chosen for the background color into a wide bowl, making sure it is well stirred and of an even consistency. If the slip is too thick, it will create visible runs on the face of the tile, and if too thin, it will leave the edges of the tile bare. Coat the best face of each tile by skimming it through the surface of the slip. Angle the tiles away from you as they enter the slip and toward you as they come out so that any excess slip runs off to one edge.

8 Leave the slipped tiles only long enough for the wet sheen to disappear from the surface before you start to draw a design. The background slip must still be soft enough for the trailed slip to sink in. If you are uncertain of your design, first draw it with a potter's needle. Load the second color into the slip trailer. With a flexible rubber-nosed trailer, the tip can be allowed to touch the surface as you draw, but with rigid plastic or glass nozzles, make sure the tip does not drag through the slip. Should your first attempts at trailing be unsuccessful, sponge off all the slip and try again.

Firing

Follow the program for a high biscuit firing (see page 58).

Rolling is the simplest way of making a tile with a unique handmade feel, and requires only a few low-cost pieces of equipment. Coating the tiles with liquid clay of contrasting color to the base tile creates a clay "canvas" on which you can draw with any metal point – the technique known as "sgraffito."

Rolled Tiles

Equipment & Materials

(To make approximately five 5 × 5-inch (127 × 127mm) tiles)

- *1 pair ⅝-inch (15mm) thick strips of wood*
- *Wooden base board*
- *Drill with standard bit*
- *4 ½-inch (15mm) metal pins*
- *2 canvas cloths*
- *4½ pounds (2kg) red terracotta clay (20% grog)*
- *Large rolling pin*
- *1 pair ⅜-inch (10mm) thick strips of wood*
- *5 small kiln batts or unglazed industrial tiles*
- *Piece of stiff plastic*
- *Potter's kidney*
- *Sheet of cardboard (or thin plywood)*
- *½-inch (15mm) brass strip (optional)*
- *Drawer handle (optional)*
- *Sharp knife or potter's needle*
- *White decorating slip*
- *Shallow bowl and pencil*
- *Small metal blade*
- *Soft brush*
- *Turquoise or colorless transparent earthenware glaze (see page 141)*

Images can also be created freehand or traced directly onto the clay using a pierced drawing (a "spons") with charcoal dust. After biscuit firing, the designs can be covered with a richly colored transparent glaze to produce tiles that are evocative of early Islamic pottery where dark figures swim below a turquoise glass.

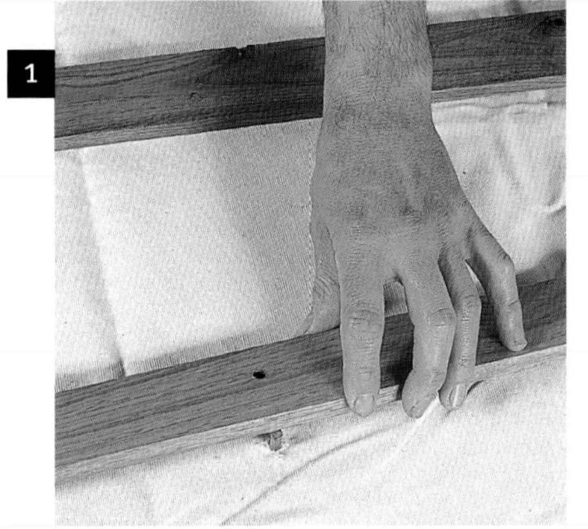

1 Position the thicker pair of wooden strips across the board, parallel to each other and spaced 2 inches (50mm) farther apart than the width of the final cut tile you intend to make. Drill through the wood into the base board to a depth of ½ inch (15mm). Place locating pins in these holes and remove the strips. Cover the

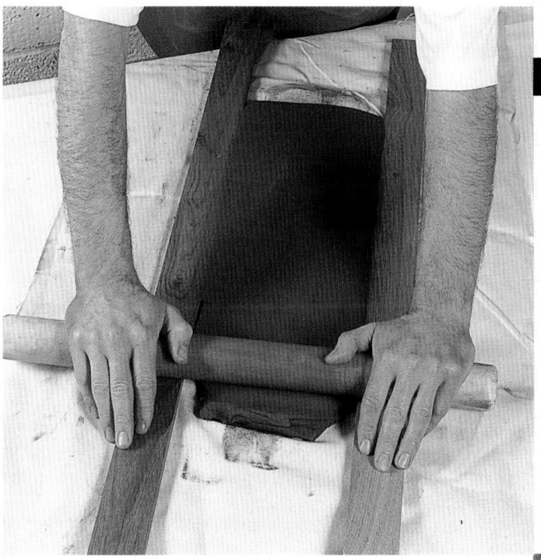

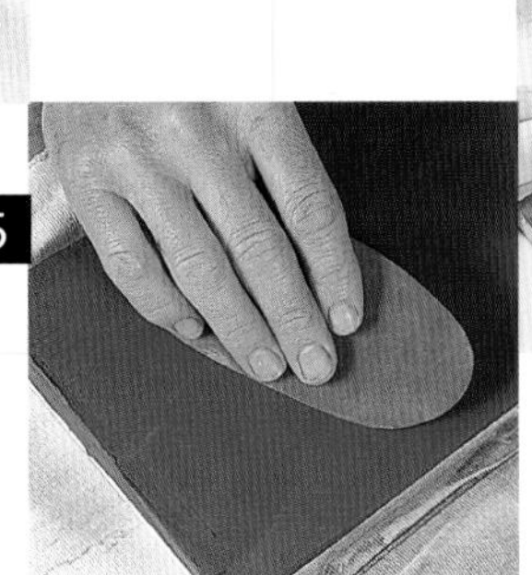

board with a smooth sheet of canvas, piercing holes in it to allow the pins to protrude through. Replace the wooden strips over the pins and check the pins are not sticking out.

2 Cut the clay from a fresh bag and place it between the strips. Using the heel of your hand, pad out the clay to elongate it in the direction you are going to roll the slab. Turn the slab over at least once during this process. Roll the slab, running the rolling pin along the strips. When the slab is level with the top of these strips, replace them with the thinner pair and continue to roll until the clay is again level with the top.

3 Remove the strips and place on top of the slab some small kiln batts or unglazed industrial tiles. These absorbent plates will be used to stack the tiles while they dry, so they must be lightweight and at least $\frac{3}{4}$ inch (20mm) wider than the tile you are cutting. Run a sharp knife around the outside edge of the plates (but do not cut out separate tiles) and trim away any excess clay.

4 Fold a flap of the underlying canvas tightly over the top of the slab and plates and turn the whole package over, leaving the clay slab on top.

5 Use the plastic to smooth out any irregularities and fill in any air pockets in the surface of the slab. Then smooth over the entire surface with a potter's soft rubber kidney. Do not push down too hard at the edges of the slab, as they can easily be deformed.

6 Cut a tile template from cardboard, its size allowing for shrinkage during drying and firing (see page 54), and mark your tiles on the clay with the point of a knife or a potter's needle. For marking a large number of tiles, you can make a plywood stamp. Screw to its edge a $\frac{1}{2}$-inch (15mm) brass strip to protrude $\frac{1}{8}$ inch (3mm) below the bottom of the wood. Make sure the brass is folded tightly to the corners of the ply by scoring the inside surface before bending and by screwing the strip into place as you go. A drawer handle can be glued to the back.

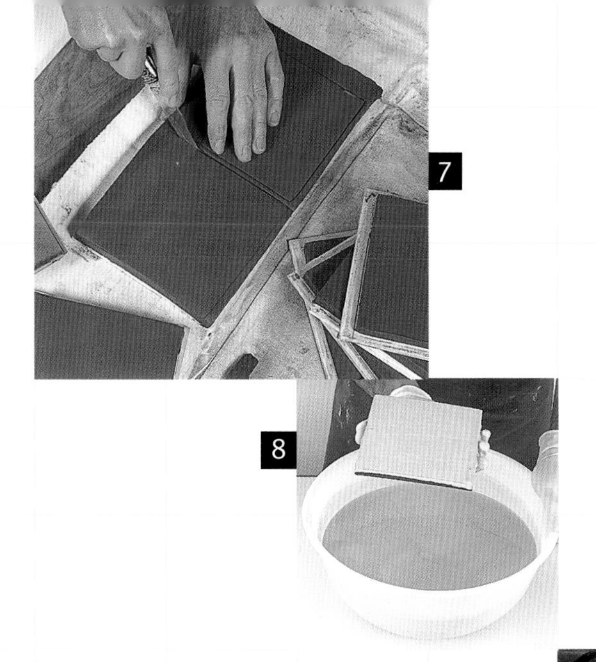

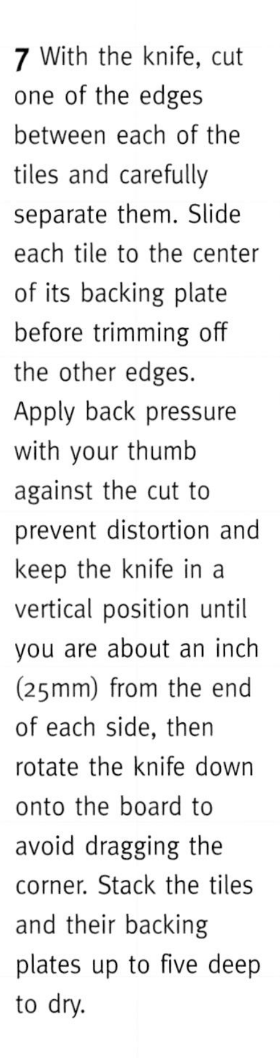

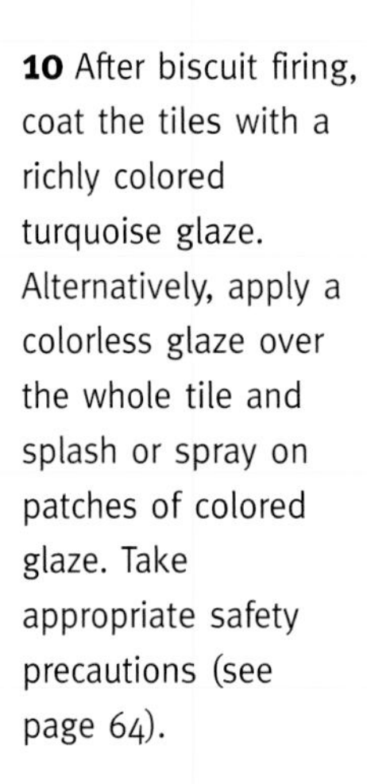

7 With the knife, cut one of the edges between each of the tiles and carefully separate them. Slide each tile to the center of its backing plate before trimming off the other edges. Apply back pressure with your thumb against the cut to prevent distortion and keep the knife in a vertical position until you are about an inch (25mm) from the end of each side, then rotate the knife down onto the board to avoid dragging the corner. Stack the tiles and their backing plates up to five deep to dry.

8 Pour the clay slip into the bowl. When the tiles are dry enough to be handled without distortion, skim them through the slip, angling them away from you as they go in and toward you as they come out so that any excess runs off. Leave the tiles to dry until the slip can be touched without causing marks.

9 Lightly sketch your design onto each tile with a soft pencil. With the small metal blade, scratch the slip back to reveal the color of the tile body beneath. Do not at this stage try to remove any small burrs of clay formed by the scratching. Leave them to dry with the tile and knock them away with a soft brush when the tile is ready to fire.

10 After biscuit firing, coat the tiles with a richly colored turquoise glaze. Alternatively, apply a colorless glaze over the whole tile and splash or spray on patches of colored glaze. Take appropriate safety precautions (see page 64).

Firing

Follow the programs for (1) low biscuit firing, (2) earthenware glaze firing (see page 58).

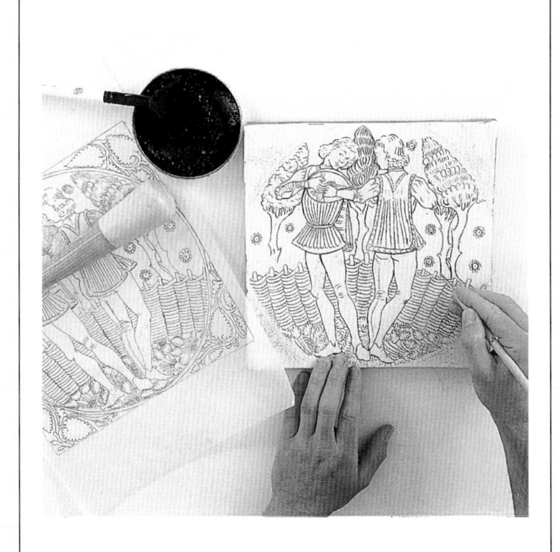

USING A "SPONS"

For reproducing large or complicated designs you can create a "spons." Copy the pattern on good-quality tracing paper, then pierce a series of holes along every line with a compass point or a large-gauge needle. Hold the spons over the tile and pounce finely ground charcoal powder through the holes with a soft brush.

Rolling allows you to produce tiles of virtually any thickness and size with the same basic equipment used in the previous project (see page 70). With various colored clays and cardboard templates, you can create an interlocking pavement with a medieval pattern.

Rolled Floor Tiles

Equipment & Materials

2 canvas cloths
Wooden base board
Buff Raku clay (40% grog)
Red terracotta clay (40% grog)
Wooden mallet and frame
Large rolling pin
Sharp knife
Absorbent wooden board
Piece of stiff plastic
Potter's kidney
Pencil and tracing paper
Carbon paper
Sheet of cardboard
Craft knife
Linseed oil

Two colors of heavily grogged clay are normally required, but if only one is available, colored slips can be applied to the tile surface to bring out the pattern. The clay must have a grog or sand content of at least 40% (see page 53). If two different clays are used, check with the manufacturers that the overall shrinkage from plastic to a high biscuit temperature of 2,012°F (1,100°C) is the same for both.

1 Lay a sheet of canvas over the wooden base board. Cut a 4½-pound (2kg) lump of clay from a fresh bag and place it on the canvas. Using your weight through the heel of your palm, move your hand around the lump, forcing the clay to spread out in all directions. Turn the slab over and repeat on the other side.

1

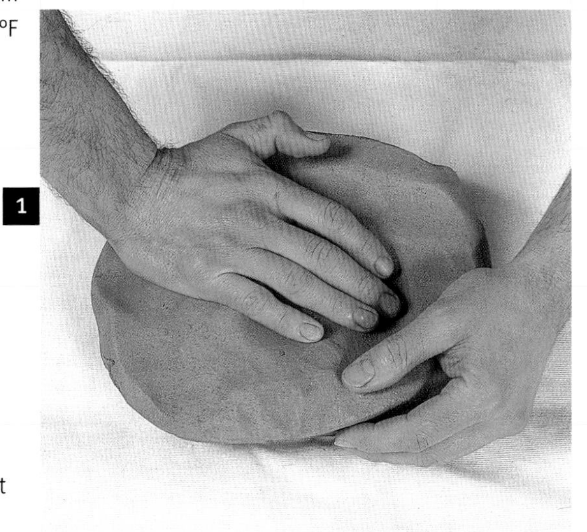

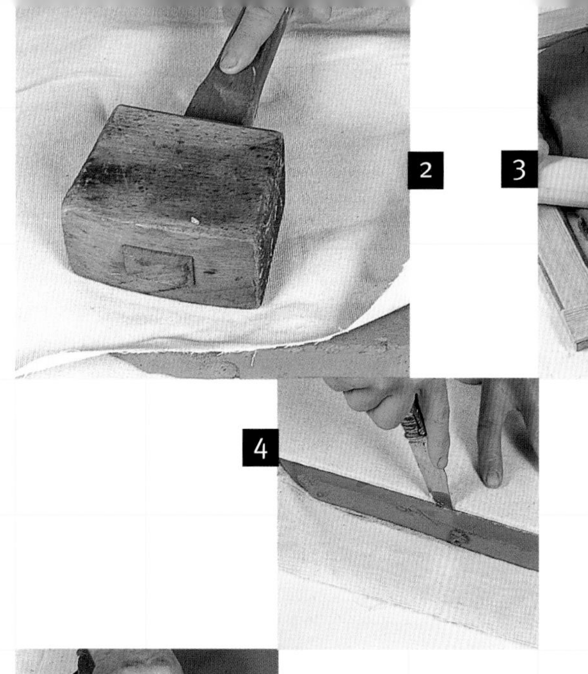

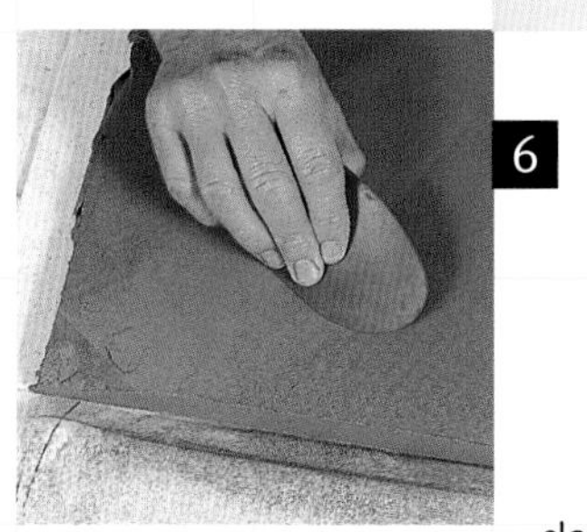

2 Complete the process by placing a second sheet of canvas on top of the clay and beating it with the flat side of the mallet. Turn the slab over and repeat.

3 Construct a simple wooden frame at least 3/4 inch (18mm) thick and 12 inches (300mm) square. Place this frame around the slab and with a rolling pin roll out the clay to fill as much of the area as possible. Tamp down any hard ridges that form in front of the rolling pin as you work. If any excess clay pushes up to the frame edge, slice it away with a knife. It can be added back onto the slab if there is a shortage of clay in any other direction, but make sure it overlaps the area and is pressed firmly into place without trapping any air.

4 Once the clay is level with the frame, remove it and place an absorbent board on top of the slab. Trim away any excess clay beyond the edges of the board.

5 Fold a flap of the underlying canvas over the slab and the backing board and use this to turn both over together.

6 The newly exposed surface of the slab will form the finished surface of the tiles. With a rigid piece of plastic, smooth out any large irregularities and fill in any air holes. Finish off by gently stroking the slab with the potter's kidney.

7 Using a photocopier or computer scanner, enlarge to life-size a picture of a medieval interlocking floor. Trace it with a pencil on a sheet of tracing paper and then use carbon paper to transfer the outlines of the individual shape elements onto a piece of cardboard. Carefully cut these out with a sharp craft knife for use as templates.

8 On the freshly rolled slab, mark the shapes of the interlocking pieces before the clay has begun to dry and shrink. Using a sharp knife, cut around the marked pieces, but do not peel away the excess clay. If there are pointed narrow corners on the shapes, do not cut completely through to the board beneath initially. Only complete such cutting when the clay has stiffened to the consistency of hard cheese. Leave pieces with tightly curved surfaces only roughly cut out until this "hard cheese" stage is reached. They can then be pared back to the exact shape with an action similar to peeling an apple.

9 After firing, the tiles simply need to be coated with linseed oil to seal the surface.

Firing

Follow the program for high biscuit firing (see page 58).

In this project, medieval tile-making techniques are adapted to modern materials and used to recreate the "encaustic" tiles still seen surviving on the floors of now-ruined European abbeys. These tiles were decorated by inlaying a colored clay into a stamped design and depicted a variety of themes.

Open-face Molded Tiles

Equipment & Materials

(To make approximately six 8 × 8-inch (200 × 200mm) tiles)

- *Wooden frame*
- *28-pound (12.5kg) bag red terracotta clay (40% grog)*
- *Potter's cutting wire*
- *Small strip of wood*
- *Potter's kidney*
- *Absorbent board*
- *8 × 8 inch (210 × 210mm) plywood*
- *Drawer handle*
- *Contact cement*
- *Photocopied image*
- *Pencil*
- *Tracing and carbon paper*
- *¼-inch (6mm) thick composite board or plywood*
- *Jigsaw with scrollwork blade*
- *1 sq yard (1sq m) ½-inch (12mm) plywood*
- *Drill with standard bit*
- *Wood hardener*
- *2 long locating pins*
- *Wooden mallet*
- *Slip trailer*
- *White decorating slip*
- *Icing side-scraper*
- *Dust mask*
- *Linseed oil or Aventurine glaze or colorless earthenware glaze*

A simple frame mold is used to make the base tiles, and a wooden stamp is used to impress a design into the clay surface. The medieval tile-maker would have made a finely tempered clay paste rich in sand and then slopped it into a wooden former or mold. A tile with similar rugged and rustic character can be made from plastic clay if it is thrown with some force into an open-face mold. Traditionally, a stamp would have been meticulously carved from a block of wood for impressing a large batch of tiles, but with modern materials several different stamps can be readily cut and assembled.

1 Assemble the frame (see page 139). Pull off small lumps of clay from a fresh bag and throw them into the frame, aiming to fill the corners first. Gradually fill in the

2

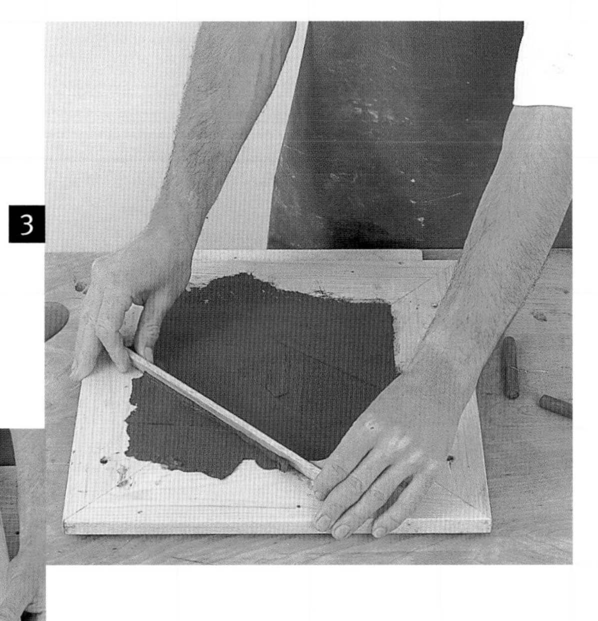

3

4

5

whole frame, working toward the center. Each lump should be thrown with enough force to deform it into the available space and expel any trapped air.

2 Pull a cutting wire across the top of the frame to trim away the excess clay.

3 Level the clay surface by dragging a strip of wood across the top of the frame, working from the center toward each corner. This surface will be the decorated face of the tile, so smooth it with a soft rubber kidney to push as much of the grog beneath the surface as possible. Use long smooth strokes, making sure the edge of the tool does not dig into the surface.

4 Lift the frame off the locating pins and invert it, still containing the tile, onto a piece of absorbent board. Cut an 8 × 8-inch (210 × 210mm) square of plywood, glue a drawer handle to the back with contact cement, and use it to push the tile out of the frame. Leave the clay to dry until it is firm enough to be handled without distorting.

5 Choose a simple medieval motif, such as the rampant lion illustrated here, and enlarge it on a photocopier to the required size. Copy the design with a pencil on a sheet of tracing paper and transfer it with carbon paper onto the sheet of composite board or plywood. Carefully cut out the pieces of the design with the jigsaw. Cut the ½-inch (12mm) plywood to 12½ × 12½ inches (320 × 320mm) and, using the frame as a drill guide, make 4 holes to align with the locating holes in the frame. With a pencil, mark the frame's edges onto this backing board and outline each element of the design within it. Using contact cement, glue the pieces of the stamp onto the board. When the adhesive has dried, coat the raised surface of the stamp with a wood hardener.

6

7

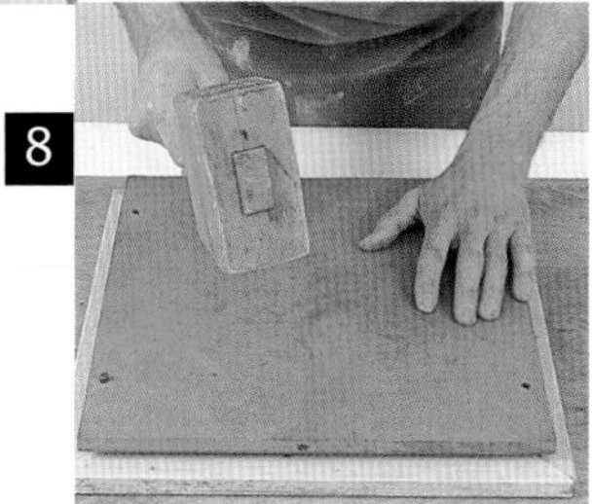
8

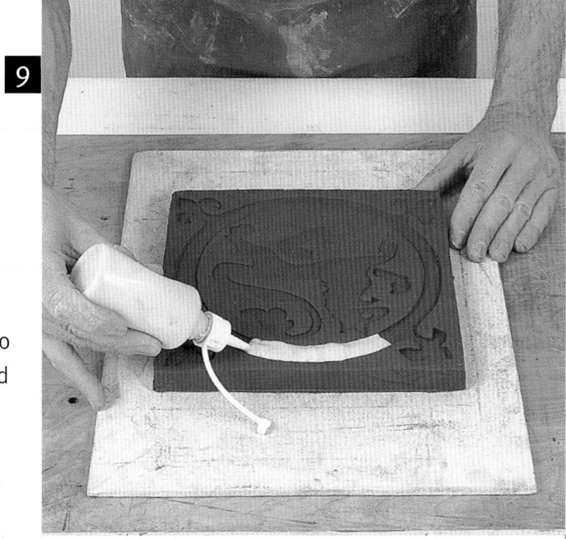
9

10

6 Relocate the tile-molding frame on its own base and place the firmed-up tile within it. Replace two of the short locating pins with longer ones that extend at least ½ inch (10mm) above the face of the frame.

7 Align the stamp over the tile, using the two long pins to register it in the correct position.

8 Once the stamp is in place, use the mallet to impress it into the tile. Work around the center of the design, hammering each position several times and applying extra blows where there are large raised areas of stamp. Avoid hammering near the edges of the backing board, as this will cause the stamp to rock. Carefully remove the stamp to reveal the impressed design. If any areas have not imprinted, you can replace the stamp to try for improvement but this will invariably blur the edges of the motif.

9 Load the slip trailer with the decorating slip (see page 63) and gradually fill in the impressed design. Let the slip flood into the depressions until it is at the point of brimming over. As it dries, it will contract and sink down, so topping up may be necessary.

10 Let the tile dry until it is just past the "leather-hard" stage, then with an icing side-scraper slowly and meticulously scrape back any excess slip to reveal the clean lines of the imprint. **Always remember when fettling dry clay to wear an appropriate dust mask** (see page 55). After biscuit firing, the tiles can either simply be treated with linseed oil, or glazed with a low-firing clear or colored transparent glaze.

Firing

Follow the programs for (1) low biscuit firing, (2) earthenware glaze firing (see page 58).

Equipment & Materials

(To make approximately 12 6 × 8-inch (150 × 200mm) tiles)

6 rolls toilet paper
2-gallon (10-liter) plastic bucket
Metal rod or electric drill with mixer-blade
Wire strainer
1 gallon (5 liters) porcelain slip
Piece of ½-inch (12mm) thick composite board
Vegetable oil
Sheet of glass
Soft soap and sponge
Smooth potter's clay
Piece of plastic
4 mold boards (see page 51)
Pencil and ruler
Small soft brush
26 pounds (12kg) pottery plaster
Surform
Pitcher
Plastic plasterer's trowel
Thin strip of wood
Potter's kidney or icing side-scraper
Sheet of cardboard
Sharp knife
Piece of silica board

Paper-clay is made by adding paper pulp to a clay slip. It is a new and versatile material which will enable you to make tiles of almost any size, with a much smaller risk of warping than any normal clay body. It will produce a surface quality similar to that of handmade rag paper.

Making Tiles with Paper-clay

Paper-clay combines the qualities of the two raw materials so that very thin sheets can be used with an etching press. Its strength and flexibility before firing allow you to join thick slabs to thin slabs and wet slabs to dry slabs. In this project you will make rectangular base tiles in an open plaster mold, using a piece of composite board as a model, and decorate them using a quick and simple printing method that is ideal for reproducing bold designs. During drying and firing, paper-clay will shrink up to 15%. Aim to cast each slab of paper-clay large enough to cut into four individual tiles. Allow at least a 1-inch (25mm) border to the slab that can be cut away and discarded if the edge is damaged when removing it from the mold.

1 Prepare the paper pulp by adding torn-up sheets of the toilet paper to warm water in a plastic bucket. You can use as many as 6 rolls of paper to every gallon (5 liters) of water depending on the absorbency of the tissue. After all the paper has been added, leave the mixture to stand for an hour, making sure there is still an excess of water. Rapidly agitate the mixture with a metal rod or an electric drill mixer-blade to speed up the disintegration of the paper **(take care not to get any water onto electrical appliances).**

1

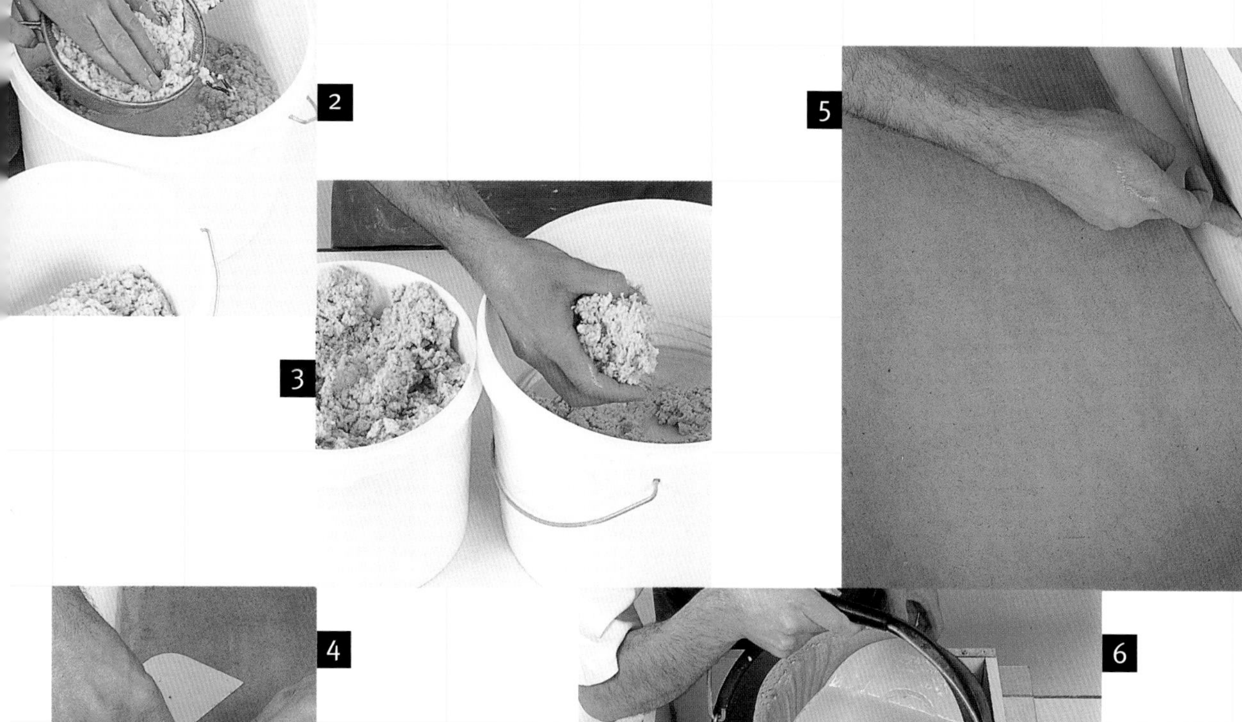

2 Leave the pulp for 24 hours, then scoop it out of the bucket with a wire strainer and squeeze out any unabsorbed water with the palm of your hand. The squeezed pulp can be stored in a freezer until required. At room temperature both paper pulp by itself and the paper-clay mixture will start to putrefy after a couple of weeks.

3 Pour the porcelain casting slip into a 2-gallon (10-liter) bucket and place a ruler down the side of the bucket to measure the depth of the fluid. Add the paper pulp until the level is approximately one and a third times the original measurement. This results in a mixture of 25% pulp to 75% clay. Thoroughly stir it and leave to stand for 24 hours. A final rapid stir will result in a paper-clay slurry with the consistency of oatmeal.

4 Next, prepare the model for the tile. Coat the board with vegetable oil to make it less absorbent and stick it down on the sheet of glass, using a thick application of soft soap to create a suction. Roll out thin coils of the potter's clay and apply to the edge of the board, thumbing it down firmly. This prevents plaster from seeping under the board and produces a beveled edge to make sure the cast-up sheets of paper-clay will drop cleanly out of the mold. To create a neat, even bevel, gradually pare away the excess clay with a piece of plastic, taking care not to apply too much pressure.

5 Place the mold boards around the model, leaving a clear border 1½ inches (40mm) wide. Apply more thin clay coils to the inside edges of the mold boards where they butt up to each other and where they touch the glass to prevent plaster from seeping out before it sets. Make pencil marks on the inside of the boards 1½ inches (40mm) above the top surface of the model to indicate the level to which plaster will be poured. Apply three thin coats of soft soap to the exposed surface of the model and the inside of the mold boards. Put each coat on with a brush and wipe it off with a clean damp sponge.

6 Make and pour the plaster (see page 65). If two pours are required, make sure the first covers the model completely. Before the plaster sets after each pour, either vibrate the table or, wearing a rubber glove, gently rub the surface of the model to release any trapped air bubbles.

7

8

9

10

11

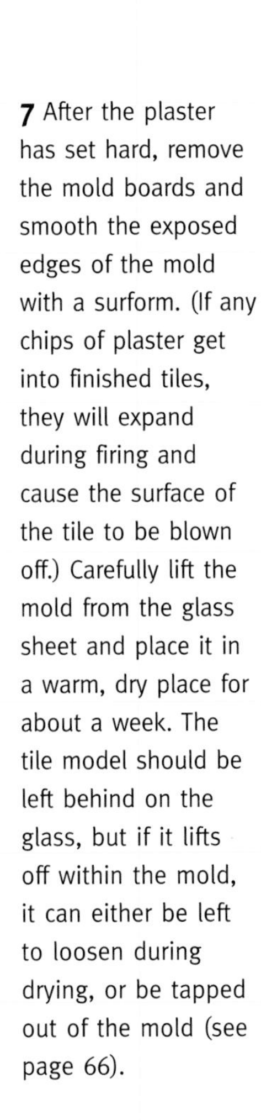

7 After the plaster has set hard, remove the mold boards and smooth the exposed edges of the mold with a surform. (If any chips of plaster get into finished tiles, they will expand during firing and cause the surface of the tile to be blown off.) Carefully lift the mold from the glass sheet and place it in a warm, dry place for about a week. The tile model should be left behind on the glass, but if it lifts off within the mold, it can either be left to loosen during drying, or be tapped out of the mold (see page 66).

8 After a week the mold will look lighter and feel warm to touch, which indicates it is ready to use for casting. Lay it on a flat table and check the inside surface for any loose pieces of plaster. Using a pitcher as a scoop, place enough paper-clay slurry to fill the mold in its center.

9 Distribute the slurry around the mold using a plasterer's trowel. Puddle it into the corners of the mold, taking care not to chip the plaster.

10 Once the pulp is evenly distributed, tamp the surface with a thin strip of wood, moving any excess slurry off to one end of the mold for recycling. This surface forms the finished surface of the tiles.

11 As the slab dries, smooth it with the rubber kidney or an icing side-scraper, but keep some texture as this gives the material its unique quality when fired. Leave the slab to dry until the edges begin to shrink away from the mold. Place a board on top of the mold and turn them over together. A gentle tap on the back of the mold should release the slab, but if not, leave the mold inverted with its edges raised on two thin strips of wood until the slab dries enough to drop out.

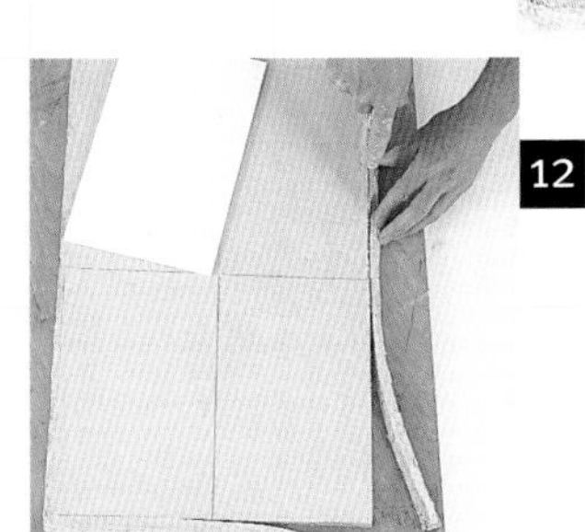

12

12 Make a template from cardboard and use it to mark your tiles on the slab. Cut the slab into tiles with a sharp knife. If it has dried too much to cut easily, score the top surface and snap the slab over the edge of a table. Leave the tiles on a piece of silica board to dry. No turning is necessary since the fibrous open texture of the paper-clay will ensure even drying. For decorating your tiles, follow the project on pages 81–3.

An ancient and widely used method of decorating pottery was to mold small bas-reliefs or medallions of clay and apply them to the surface of a vessel. These clay motifs, known as "sprigs," were usually formed by pressing plastic clay into a carved mold.

Decorating Paper-clay Tiles

Equipment & Materials

(To decorate 12 prepared paper-clay tiles)

Smooth potter's clay
Sheet of glass
Plastic food container
3 pounds (1.5kg) plaster
Soft soap
Pencil
Drawing paper
Carbon paper
Sgraffito or linoleum-cutting tool (see page 50)
Porcelain casting slip
Paper-clay mixture (see page 78)
Decorating slip in color of choice
Prepared paper-clay tiles
Piece of plastic or an icing side-scraper
Absorbent board
Wax emulsion
Latex resist
White matte earthenware glaze
Wide bowl or container
Small knife or spatula
Dust mask
Underglaze powdered pigments
Water-based underglaze medium
Fine paintbrush
Colorless transparent earthenware glaze
Slip trailer or broad soft brush

This simple method of relief "printing" can be readily adapted to reproduce a lively embossed design on a pad of paper-clay. These sprigs can be stuck to the base tile at virtually any stage before the biscuit firing. Designs can either be filled in, using bright and flamboyant underglaze colors, or reproduced as a simple line drawing by "inking up" the sprig mold with a colored slip.

1 To make the blank sprig mold, first form a thin clay slab as a model. Moisten its back and stick it down to a sheet of glass, making sure the edges of the slab slope gently out and that any sharp corners are rounded off. When a pressing is taken of the sprig, this will form the boss on which the design is raised. Use an old plastic food container with its base cut away as the confining frame for the plaster. Invert it over the clay model and thumb a thick roll of clay around its edge to hold it down on the glass. Soft-soap the inside of the container and pour in the prepared plaster (see page 65). When it has hardened, remove the container and peel out the clay slab.

2 Make one sprig mold for each design you wish to reproduce. With a pencil, draw the design on a sheet of paper to fit within the dimensions of the sprig and then transfer it with carbon paper to the recessed face of the mold.

3 Using a sgraffito cutting tool, engrave the lines into the plaster. The deeper they are carved, the more open and rounded in profile they must be cut. If they are narrow and straight-sided, the raised detail they form may snap off as the sprig is removed from the mold.

4 Paint decorating slip into the carved design, taking care to keep it within the engraved lines.

5 As paper-clay has a coarse fibrous texture, spread over the carved surface a small amount of porcelain casting slip used as a base for making the paper-clay tiles (see page 78) to make a cleaner reproduction from the mold. Add just enough slip to form a thin layer over the decorated face and carefully swirl it around to coat the entire inner surface. It will immediately form a thin skin of plastic clay against the absorbent surface.

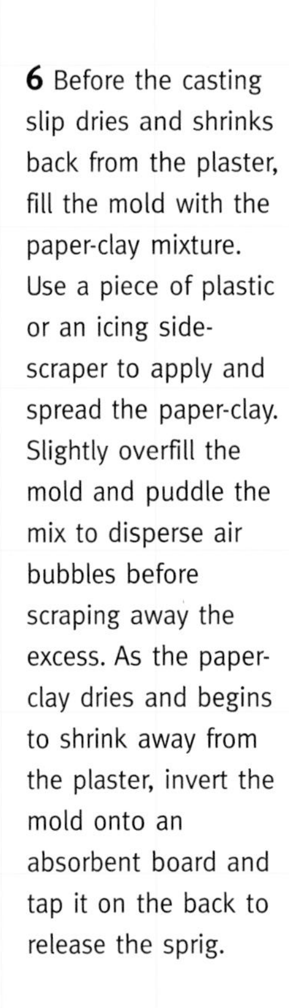

6 Before the casting slip dries and shrinks back from the plaster, fill the mold with the paper-clay mixture. Use a piece of plastic or an icing side-scraper to apply and spread the paper-clay. Slightly overfill the mold and puddle the mix to disperse air bubbles before scraping away the excess. As the paper-clay dries and begins to shrink away from the plaster, invert the mold onto an absorbent board and tap it on the back to release the sprig.

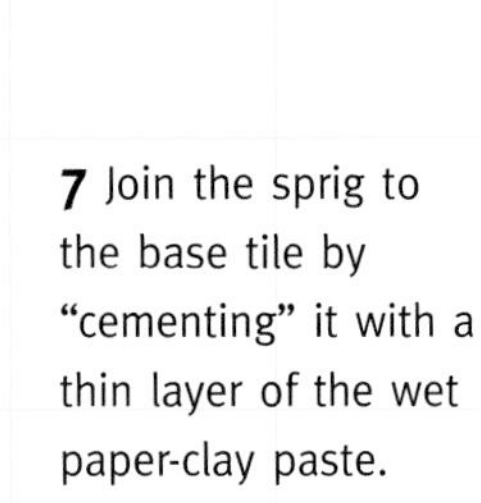

7 Join the sprig to the base tile by "cementing" it with a thin layer of the wet paper-clay paste.

8

9

10

11

12

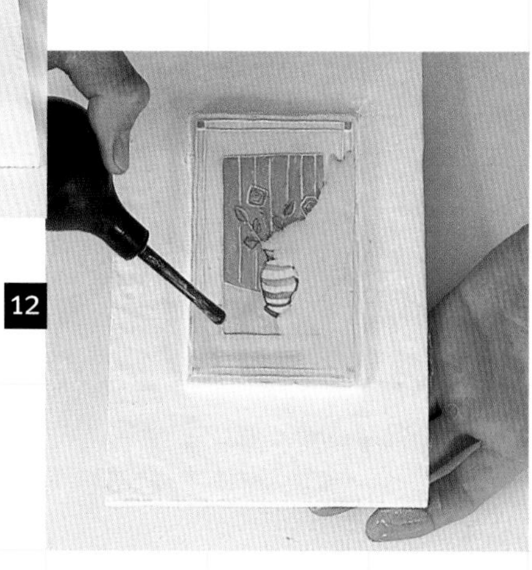

8 After biscuit firing, apply wax emulsion to the back of the tile and then paint a latex resist over the face of the sprig. This forms a rubbery mask while a background glaze is applied to the base tile and can then be peeled away. The thicker the skin, the easier it is to peel. Use an old paintbrush to apply the resist and clean it immediately in soapy water.

9 Pour the white matte glaze into a container. Immediately before dipping the tile into the glaze, first dip it into water to pre-saturate the subsurface. Otherwise, due to its high porosity, the paper-clay will absorb an overly thick layer of glaze, some of which will enter the tile body and may cause cracking and warping after firing.

10 After glazing, peel away the latex skin using a small knife or spatula to pull up one of the corners. If there are still wet beads of glaze on the surface of the latex, carefully gather the skin in from all four corners to avoid drips on the tile surface.

If you peel the latex away after the glaze is fully dry, wear a mask and work either in a spray booth or outside where any fine glaze dust will be safely dispersed.

11 Now paint the underglaze pigments onto the design with a fine brush. These glazes **should be treated as toxic and handled in accordance with the precautions recommended by the manufacturer.** The powder can be mixed with either an underglaze medium or a small amount of transparent glaze. Water-based media are preferable since the colors can be thinned with water and a glaze applied over the top as soon as they are dry.

12 Flood the colorless transparent glaze onto the sprig with a slip trailer or paint it on with a broad soft brush. A thick application will produce a rich and often crazed surface, whereas a thinner glaze will look harder and brighter.

Firing

Follow the programs for (1) high biscuit firing, (2) earthenware glaze firing (see page 58).

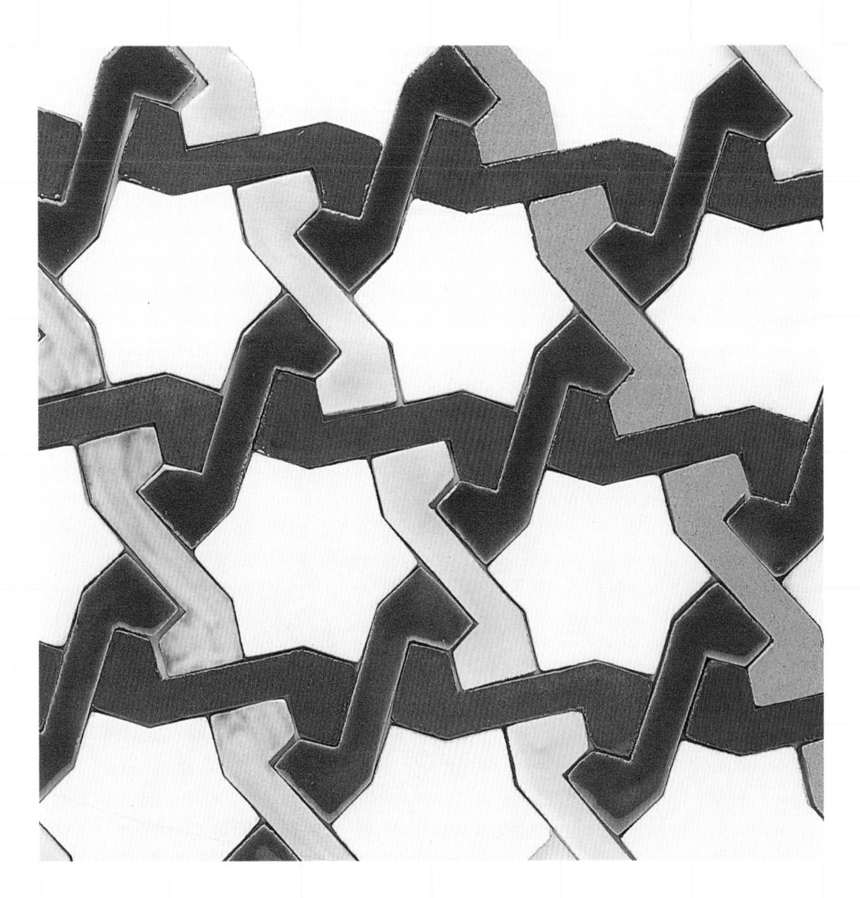

A tile extruder is one of the most versatile tools in a pottery, and most school or college workshops will have this type of machine. This project and the following Delft Design project show how it is used.

Extruded Tiles

Equipment & Materials

Copy of design
Tracing paper
Carbon paper
Sheet of cardboard
Craft knife
2 die plates
Clay extruder with 6-inch (150mm) barrel and 10-inch (250mm) expansion box
Vegetable oil
28-pound (12.5kg) bag buff clay (40% fine grog)
Wooden board
20 6 × 6-inch (150 × 150mm) square tile batts or biscuit tiles
Wooden frame
Potter's clay-cutting harp
2 absorbent boards
Turquoise transparent earthenware glaze
White matte glaze
Opaque green glaze
Aventurine glaze
Protective gloves
Shallow bowl

Basically, an extruder forces clay through a steel die (a plate with an opening cut in it) to make geometric tiles of almost any shape. A local metalworker will be able to cut an opening to suit your design if you do not have the right tools. In this project, two simple shapes and four colored glazes are used to echo elements of the interlocking tiles (tesserae) seen in ancient Moorish palaces.

1 Trace your design and transfer it with carbon paper to a piece of cardboard. Cut out the shapes with a craft knife and carefully draw around each one onto the center of a blank die plate. Check that the resulting opening will occupy no more than a quarter of the total die plate area before cutting it out. If it is larger, it is unlikely that the clay will extrude evenly.

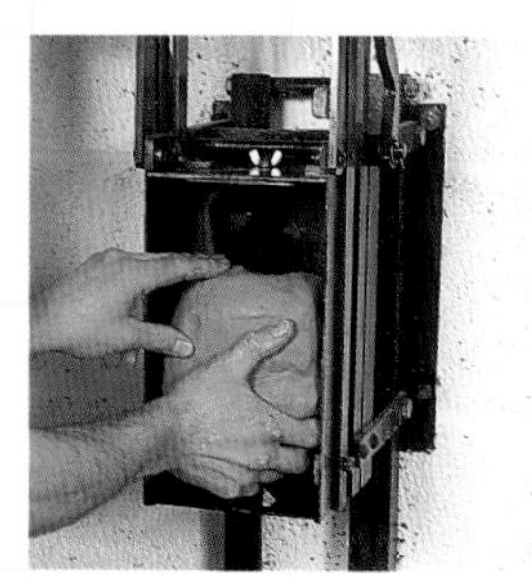

3

4

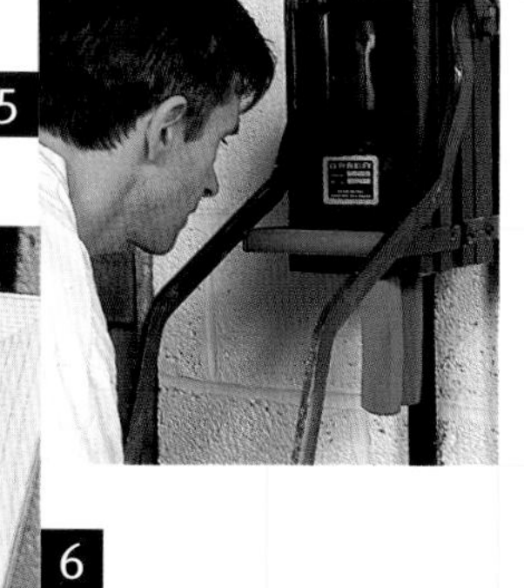

5

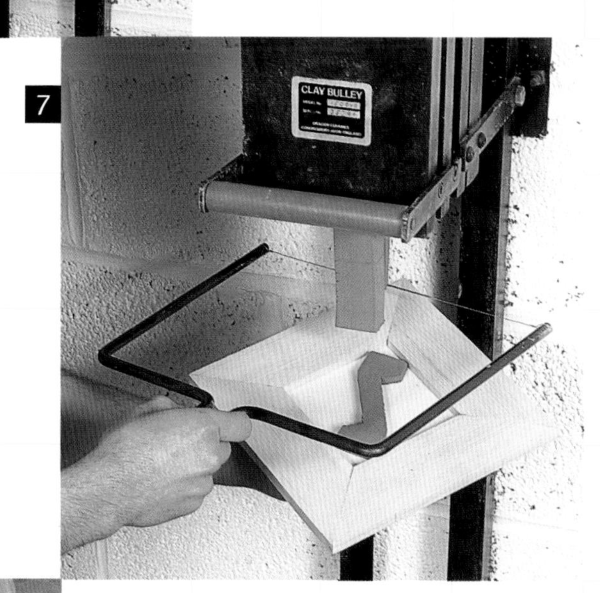

7

6

8

9

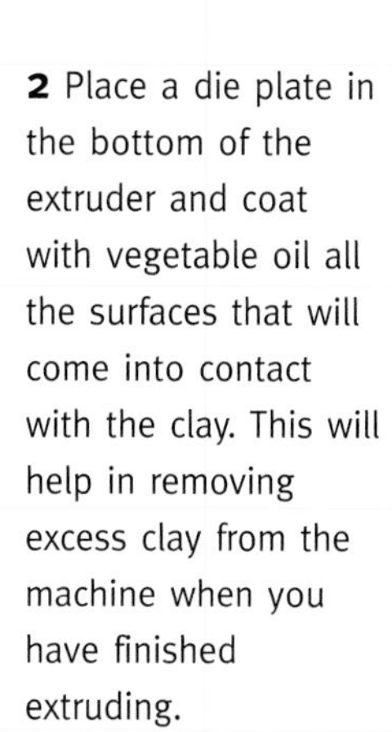

2 Place a die plate in the bottom of the extruder and coat with vegetable oil all the surfaces that will come into contact with the clay. This will help in removing excess clay from the machine when you have finished extruding.

3 Cut a lump of clay from a new bag and, by repeatedly dropping it onto the wooden board, shape it into a block that will fit into the barrel of the extruder.

4 Lift the piston out of the barrel and load the clay. All extruders can be loaded from the top, but in some machines the front of the barrel can also be opened to allow a large chunk of clay to be placed inside.

5 When the chamber is fully loaded, push the piston back into the barrel by pulling down on the lever. This forces the clay through the die.

6 Take a biscuit tile or a small kiln batt and make a wooden frame that will sit securely over it, holding the batt in a recess. This frame acts as a cutting guide, and its thickness will determine that of the tesserae, which should be made at least $^3/_8$ inch (8mm) thick to lessen the chances of distortion during drying.

7 Bring the framed tile up to the bottom of the extrusion. Holding it firmly up against the clay face, slide the clay-cutting harp across the upper surface of the wooden frame, slicing off a single tessera.

8 Keep on cutting off tesserae until the whole extrusion is used before recharging the barrel with fresh clay. Pack the cut pieces closely

together on absorbent boards or biscuit tiles to dry.

9 After biscuit firing, try various glazes on the two elements, experimenting with combinations of color and texture. Wearing protective gloves, pour the glaze into a shallow bowl, keeping the glaze "slop" thick, and carefully skim the pieces through its surface.

Firing

Follow the programs for (1) low biscuit firing, (2) earthenware glaze firing (see page 58).

The beauty of Delft tiling relies on a background network of interlocking corner motifs. Within this net float images of animals, plants, people, and landscapes. These designs are reproduced on biscuit-fired tiles with charcoal pushed through a pierced drawing known as a "spons," and then painted.

Delft Design

Equipment & Materials

(To make approximately 25 5 × 5-inch (127 × 127mm) tiles)

- *Comb die plate*
- *Wall-mounted clay extruder with 6-inch (150mm) barrel and 10-inch (250mm) expansion box*
- *28-pound (12.5kg) bag red terracotta clay (20% grog)*
- *5 6-inch (150mm) wide absorbent soffit boards*
- *Potter's clay-cutting harp*
- *Piece of plastic or icing side-scraper*
- *Potter's kidney*
- *Wooden cutting board*
- *Steel chisel*
- *Wax emulsion*
- *Tin glaze*
- *Copy of Delft design*
- *Tracing paper*
- *Charcoal sticks*
- *Soft brush*
- *Fine paintbrush*
- *Lining pigment (see page 141)*
- *Lath of wood*
- *2 strips of wood*
- *Roundheaded paintbrush*

For the corner motifs to align with one another, the tiles must be accurately cut square and remain relatively unwarped during drying and firing. The Delft potters used a very sandy clay pressed into open-face molds but, in a small studio, extrusion offers a quick and easy method for producing high-quality tiles with the weight and feel of the traditional product. **This project requires some assistance from a helper.**

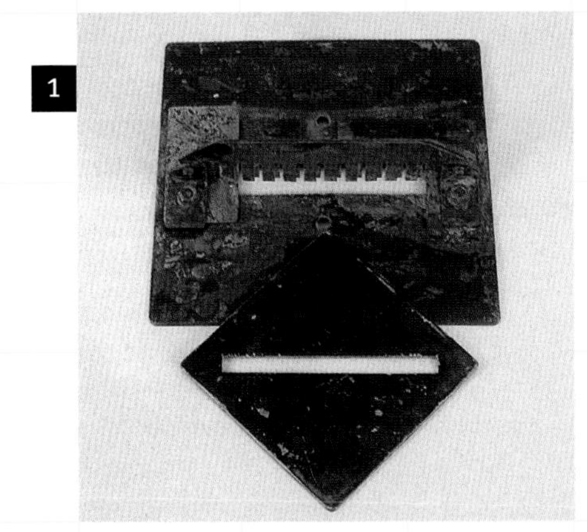

1 Two types of die can be used for extruding square wall tiles: a simple slot, the same width and thickness as the tile you wish make, or a double-thickness slot with a comb structure mounted above it. The former produces a straight-edged slab ready for cutting into individual tiles. The latter, which is used in this project, forms

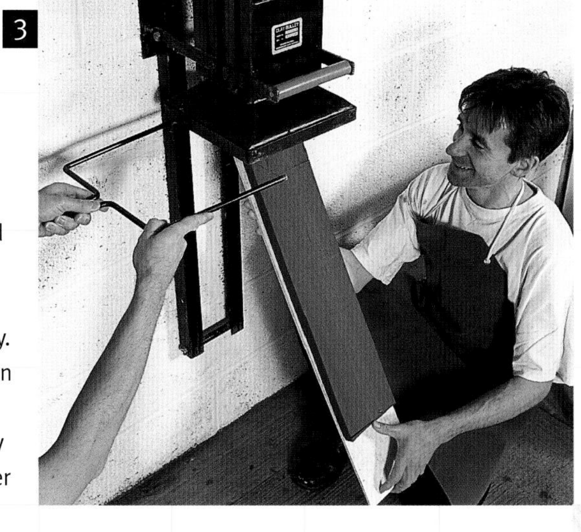

a double slab joined down the center of the extrusion by narrow ridges of clay. This double extrusion is cut up into pairs of tiles that will only be broken apart after the biscuit firing.

2 Fully load the barrel of the extruder with clay and pull down the lever. Aim to form an extrusion four and half tile-widths long. After trimming and cutting this will leave you with four pairs of tiles. Longer than this, and most clays will start to stretch and thin under the hanging weight of the extrusion. It might be necessary to reload the barrel during the extrusion. This will form a small ridge on both faces of the extrusion which must be fettled away afterward.

3 With your assistant take the piece of soffit board, which must be at least five tiles long, and hold it directly behind the extrusion. Gradually move the bottom of the board forward and up, taking the weight of the extrusion as your colleague slices it free of the machine with a clay-cutting harp. Poor timing will leave you with a crumpled mass of clay on the floor. Carefully carry the board to a drying bench. Align the edge of the extrusion with the edge of the soffit board and leave it for two or three hours to stiffen.

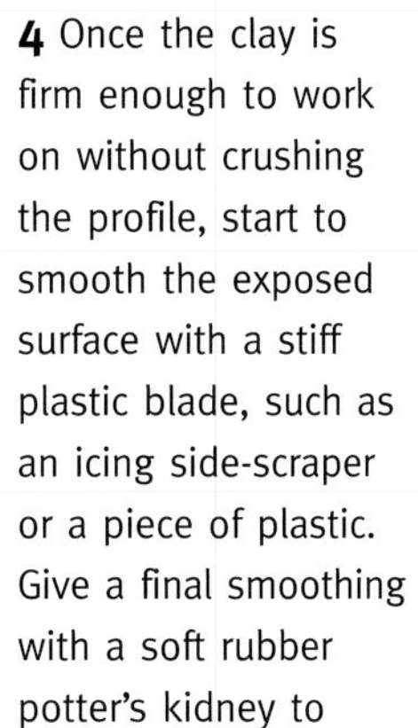

4 Once the clay is firm enough to work on without crushing the profile, start to smooth the exposed surface with a stiff plastic blade, such as an icing side-scraper or a piece of plastic. Give a final smoothing with a soft rubber potter's kidney to push any grog beneath the surface.

5 Place the cutting board upside down over the extrusion. If there is any surface roughness on the board, place a strip of thin paper such as newsprint between the two. Make sure both extrusion and soffit board lie between the vertical walls of the cutting board. Holding everything in a firm sandwich, turn the whole assemblage over. Remove the soffit board and smooth the newly exposed surface of the extrusion. With this double tile method, both the top and bottom surfaces of the extrusion will form the face of finished tiles.

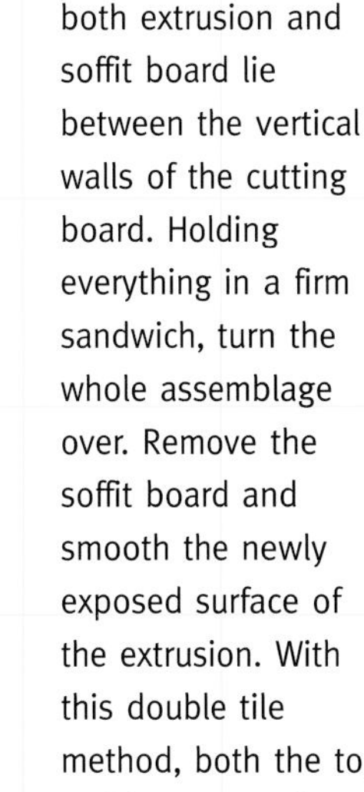

6 Measure the width of the extrusion and mark off equal lengths along one edge. Discard any clay at either end that has been disfigured by handling or has dried out.

7 Using the clay-cutting harp and the cutting guide at the end of the board, slice off the pairs of tiles from one end of the extrusion. Clean off any burls on the cut edges with the side of your fingertip. Leave the tiles to dry either standing on their edges, card-house fashion, or lying face down on the soffit board.

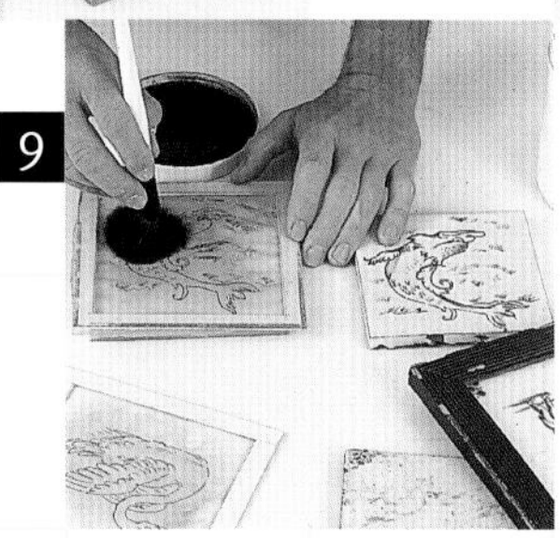

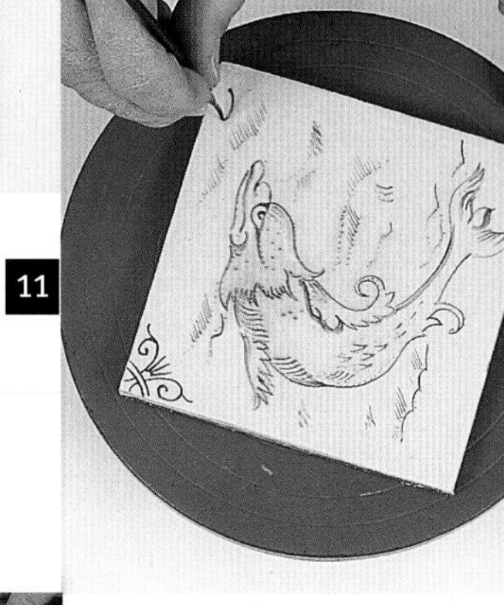

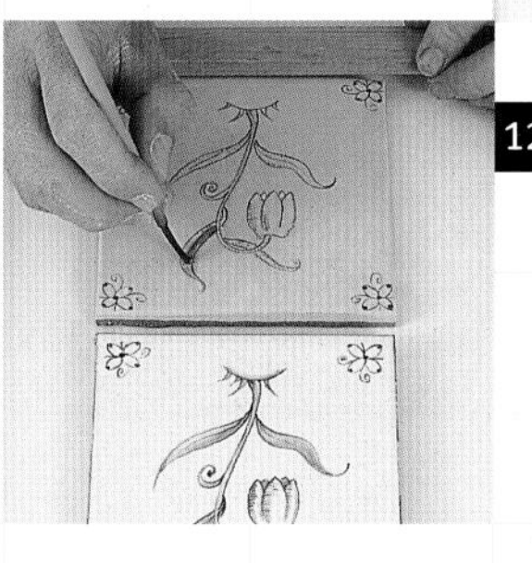

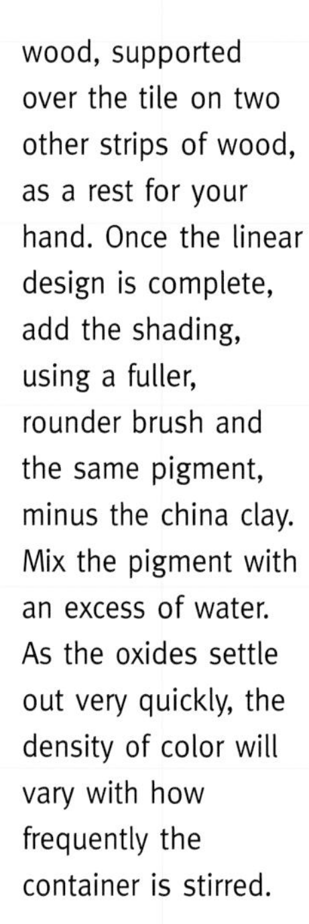

8 For the biscuit firing, stand the tiles on their edges on top of a thin layer of sand (see page 59). Placing them tightly together in an edge-to-face arrangement will avoid a domino effect if one does fall over. After firing, snap the pairs apart with a blow from a broad steel chisel, revealing the ridged profile at the center of the extrusion. Wax the back of the tiles and dip them in a tin glaze (see page 141).

9 Copy your chosen Delft design on good-quality tracing paper and make a "spons" (see page 72). If you plan to use a design regularly, it is worth mounting the spons in a cardboard frame, which will also help with positioning the design on the tile. Crush the charcoal sticks and grind them down finely with a pestle, then pounce the dust through the spons with a soft brush. The image left on the tile will act as a guide for your painting. The charcoal will burn away completely in the firing.

10 With a fine brush, paint in the linear design using a lining pigment made from cobalt and china clay (see page 141). To avoid smudging your work, place a lath of wood, supported over the tile on two other strips of wood, as a rest for your hand. Once the linear design is complete, add the shading, using a fuller, rounder brush and the same pigment, minus the china clay. Mix the pigment with an excess of water. As the oxides settle out very quickly, the density of color will vary with how frequently the container is stirred.

11 Add the corners using the china clay lining pigment. The vast variety of designs seen on the old Delft tiles are generally stylizations of only three basic types: the "Ox-head," "Fleur-de-lis," and "Spider's-head" corners. If you are painting a large number of tiles, it is worth using a turntable to rotate them from corner to corner. The head of the turntable can be marked with reference points to ensure a similar alignment of the motifs on each tile.

12 The traditional blue and white of Delft tiles can be varied with other colors. The tile illustrated has been decorated with the "Spider's-head" corner.

Firing

Follow the programs for (1) low biscuit firing, (2) earthenware glaze firing (see page 58).

Bas-relief tiles can create a rich textured surface that turns a tiled wall into a striking architectural feature. The painstaking process of repeatedly modeling relief tiles can be alleviated by the use of a simple press mold. Make one original and then reproduce as many copies as you like.

Press Molded Tiles

Equipment & Materials

- *Photocopy of design*
- *Pencil*
- *Tracing paper*
- *Carbon paper*
- *Sheet of cardboard*
- *Craft knife*
- *Rolling pin*
- *Buff clay (40% fine grog)*
- *Sharp knife*
- *Plastic wrap*
- *Smooth potter's clay*
- *Cut-away pottery tools*
- *24 × 24-inch (600 × 600mm) sheet of glass*
- *Piece of plastic or icing side-scraper*
- *6½ feet (2m) of 2½ × 1¼-inch (60 × 35mm) wooden strip (for mold boards)*
- *32 inches (800mm) of ⅜-inch (10mm) steel studding (for mold boards)*
- *Soft soap and talc*
- *28 pounds (12.5kg) pottery plaster*
- *Sheet of cloth*
- *Potter's clay-cutting wire*
- *2 laths of wood*
- *Absorbent board*
- *Vitreous slips in 3 colors*
- *Shallow bowl*

You can abandon the confines of square or geometric tiles as press molding allows you to reproduce virtually any shape. This project develops one of the inspiring patterns of Mauritis Escher, who created interlocking motifs from natural forms. Here, a scallop, two whelks, and a starfish join as four congruent figures that totally fill in the surface, capturing the essence of a shell-covered beach. Reproduce this design on a wall or floor and use the soft finish of vitreous slips for a surface with a natural feel.

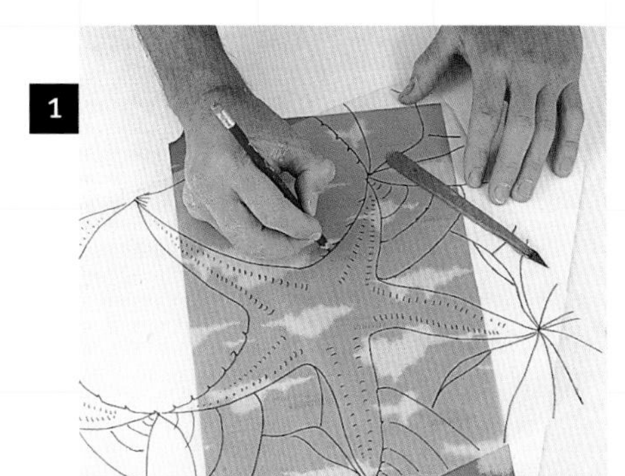

1 Referring to a source book, enlarge the design to the required size with a photocopier or computer scanner. Trace the four shapes and transfer them onto a sheet of cardboard with carbon paper. Carefully cut out each element with a sharp craft knife.

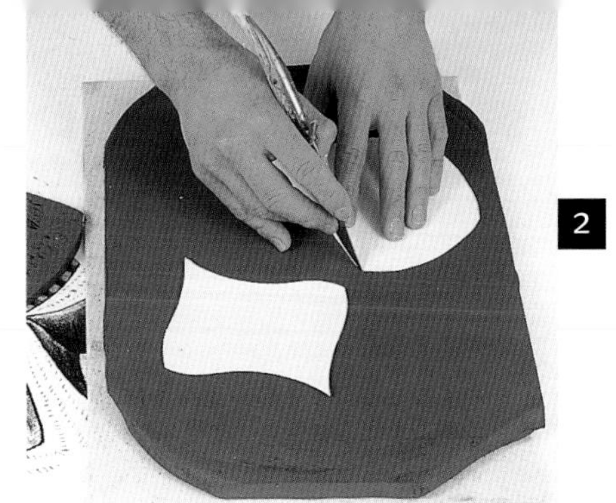

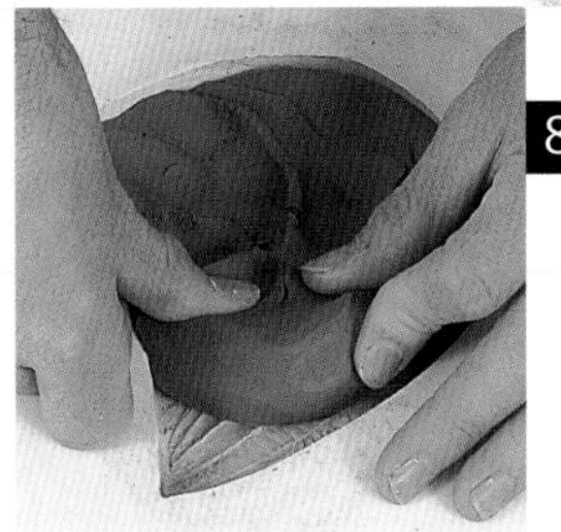

2 With a rolling pin, roll out a $3/8$-inch (10mm) thick slab of the grogged clay and use the cardboard templates to cut out the four shapes. Keep your knife vertical as you move around the profiles, making sure the templates do not slip. If you are working on a soft board, pins can be used to hold the templates in place. At this stage do not try to peel away the surrounding clay or to move the shapes, but let them stiffen in situ.

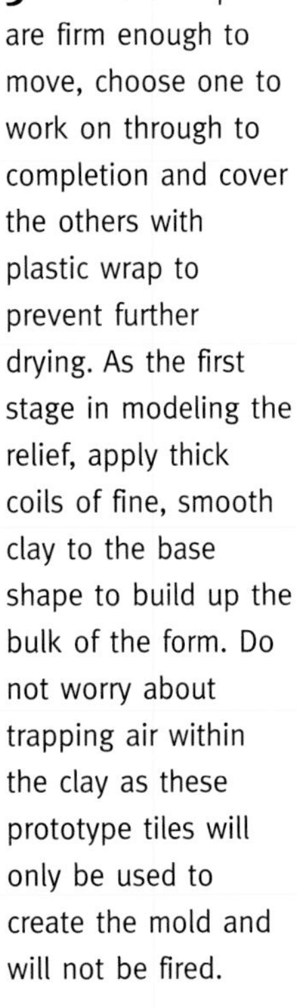

3 When the shapes are firm enough to move, choose one to work on through to completion and cover the others with plastic wrap to prevent further drying. As the first stage in modeling the relief, apply thick coils of fine, smooth clay to the base shape to build up the bulk of the form. Do not worry about trapping air within the clay as these prototype tiles will only be used to create the mold and will not be fired.

4 Continue to add the smooth clay and use a combination of carving back and modeling up to achieve the finished figure. Wire cut-away pottery tools and forged dentistry tools

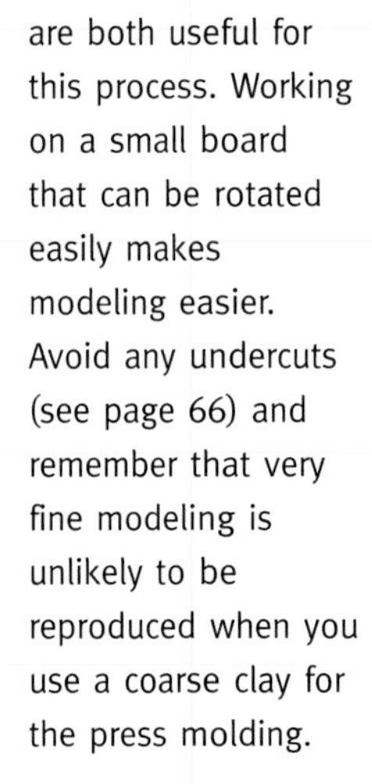

are both useful for this process. Working on a small board that can be rotated easily makes modeling easier. Avoid any undercuts (see page 66) and remember that very fine modeling is unlikely to be reproduced when you use a coarse clay for the press molding.

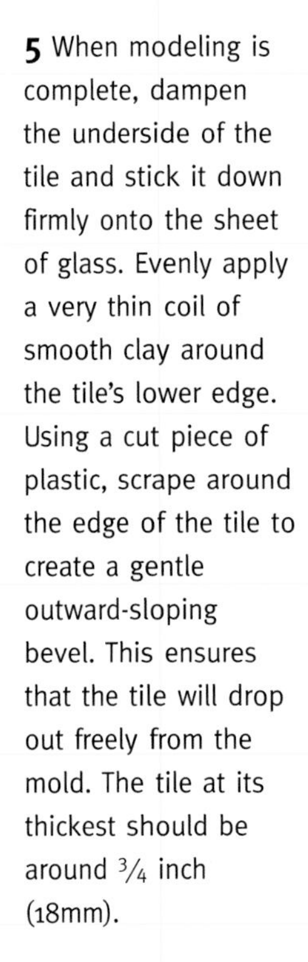

5 When modeling is complete, dampen the underside of the tile and stick it down firmly onto the sheet of glass. Evenly apply a very thin coil of smooth clay around the tile's lower edge. Using a cut piece of plastic, scrape around the edge of the tile to create a gentle outward-sloping bevel. This ensures that the tile will drop out freely from the mold. The tile at its thickest should be around $\frac{3}{4}$ inch (18mm).

6 Construct a mold frame (see panel) and center it over the model. Seal the corners of the frame and the junctions between the frame and the glass with thin coils of fine clay. Apply soft soap to the frame and glass, but not to the clay model. You are now ready to mix the plaster and pour the mold (see page 65). The plaster mold must be at least $1\frac{1}{2}$ inches (40mm) thick when dry. To help it dry quickly, pull out the model once the plaster has set hard.

7 Leave the mold to dry until it becomes whiter and warm to touch. It is now ready for use, but before you take each molding, lightly dust the inside with talc. This acts as a releasing agent which will reduce the time taken for the tile to drop out of the mold.

8 Take a ball of the fine grogged clay and press it firmly into the center of the mold. Using your thumbs, work it out into the extremities of the mold so that a continuous even layer covers the entire inside surface.

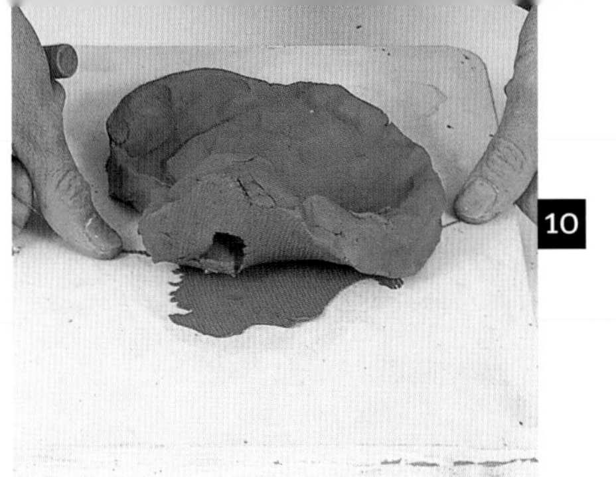

10

11

12

13

9 Place a second ball of clay in the center of the mold and press it down firmly onto the first layer. Again carefully work it out from the center to the corners, making sure no air pockets are trapped between the two layers. The mold is now filled with an excess of clay which must be compacted. Beat it several times with the side of your fist, but if clay sticks to your hand and pulls back out of the mold, cover it with a sheet of dry cloth.

10 Trim away any superfluous clay which still overlaps onto the upper surface of the mold by folding up the clay edge and pulling a potter's wire across the top.

11 Smooth off the back of the tile by pulling a piece of cut plastic or an icing side-scraper across the clay surface. Only work from the center of the mold out, turning the mold after each stroke, until the whole back is smoothed. Do not try to scrape the whole surface in a single stroke, as this pulls the clay away from the inner wall of the mold and distorts the tile's shape.

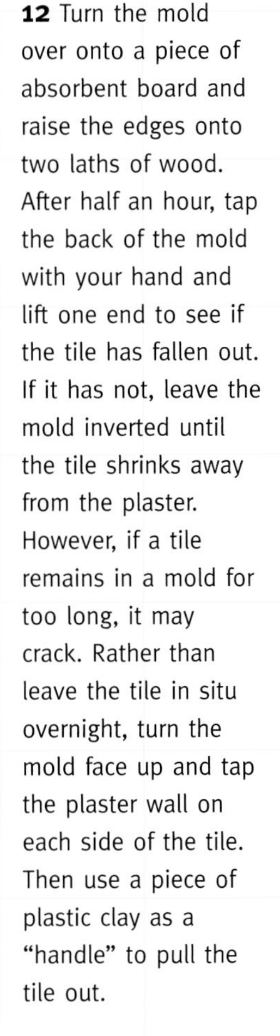

12 Turn the mold over onto a piece of absorbent board and raise the edges onto two laths of wood. After half an hour, tap the back of the mold with your hand and lift one end to see if the tile has fallen out. If it has not, leave the mold inverted until the tile shrinks away from the plaster. However, if a tile remains in a mold for too long, it may crack. Rather than leave the tile in situ overnight, turn the mold face up and tap the plaster wall on each side of the tile. Then use a piece of plastic clay as a "handle" to pull the tile out.

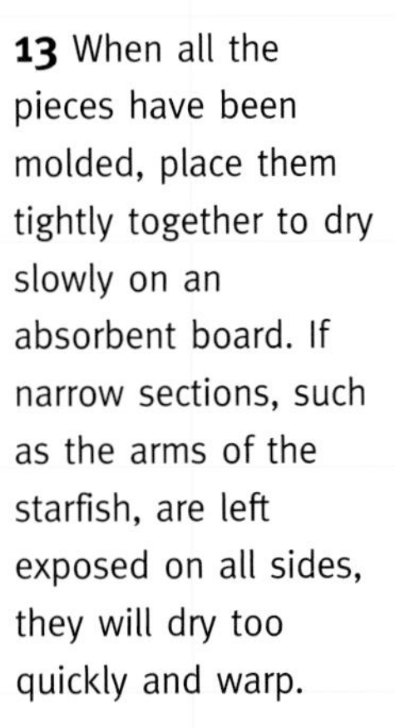

13 When all the pieces have been molded, place them tightly together to dry slowly on an absorbent board. If narrow sections, such as the arms of the starfish, are left exposed on all sides, they will dry too quickly and warp.

14

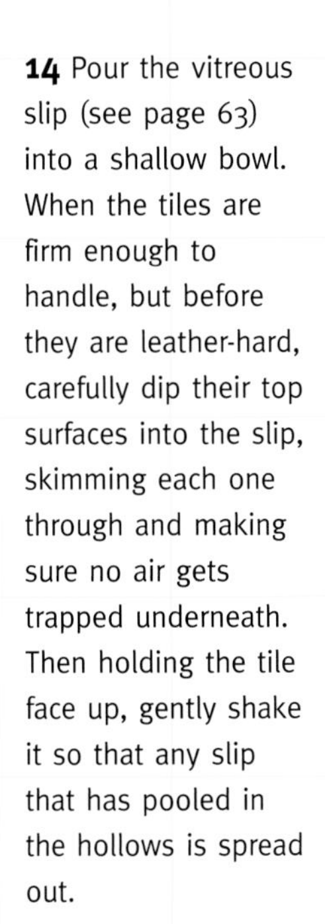

14 Pour the vitreous slip (see page 63) into a shallow bowl. When the tiles are firm enough to handle, but before they are leather-hard, carefully dip their top surfaces into the slip, skimming each one through and making sure no air gets trapped underneath. Then holding the tile face up, gently shake it so that any slip that has pooled in the hollows is spread out.

Firing

Follow the program for high biscuit firing (see page 58).

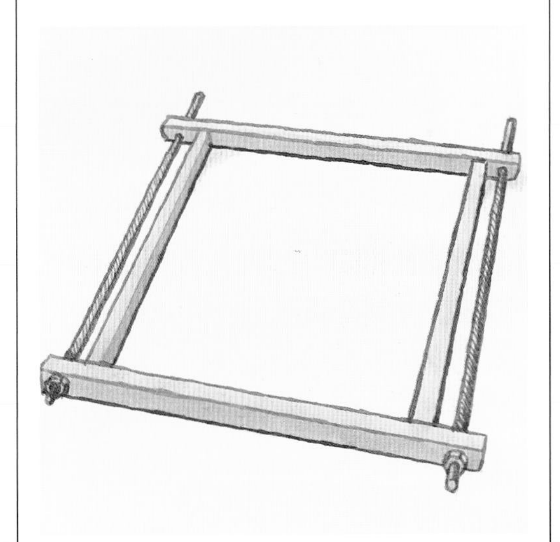

MAKING THE MOLD BOARDS

Cut two 20-inch (500mm) lengths and two 12-inch (300mm) lengths of 2½-inch (60mm) wide wooden strips. Drill a ½-inch (12mm) hole, 2 inches (50mm) in from each end of the 20-inch (500mm) lengths. The four pieces can then be clamped together into a square frame using two 16-inch (400mm) lengths of ⅜-inch (10mm) studding.

Using the same basic techniques as described in the Escher design project (see pages 89 to 91) you can create a press mold from a natural object without the initial clay modeling. An orange was chosen here, but any fruit or vegetable would look attractive.

Press Molded Tiles: Fruit Design

Equipment & Materials

Heavy cardboard
Craft knife
Canvas cloth
Rolling pin
Buff clay (40% grog)
Wooden board
Paintbrush
Fresh orange and leaf
Dressmaker's pins
2 $9\frac{3}{4}$-inch (250mm) lengths wood 2 × 2 inches (50 × 50mm)
2 6-inch (150mm) lengths wood 2 × 2 inches (50 × 50mm)
$5\frac{1}{2}$ pounds (2.5kg) pottery plaster
Surform
Tweezers
Wooden strip
Colorless transparent earthenware glaze
Glaze stains
Aerosol spraygun

Use a subject with distinctive texture and form that will be retained sharply in a relief model. Several small press molded tiles can be incorporated very effectively into a tiling scheme on a wall. For purely decorative purposes, you could set a tile into a picture frame.

1

1 On a sheet of cardboard draw a square tile $4\frac{1}{2} \times 4\frac{1}{2}$ inches (110 × 110mm) and cut it out with a sharp blade. Working on a canvas cloth, roll out a block of clay to a thin slab. Place the template on top and cut out the shape. Leave this tile model to dry to the "leather-hard" stage.

2 Transfer the tile model to a wooden board and, with the paintbrush, moisten

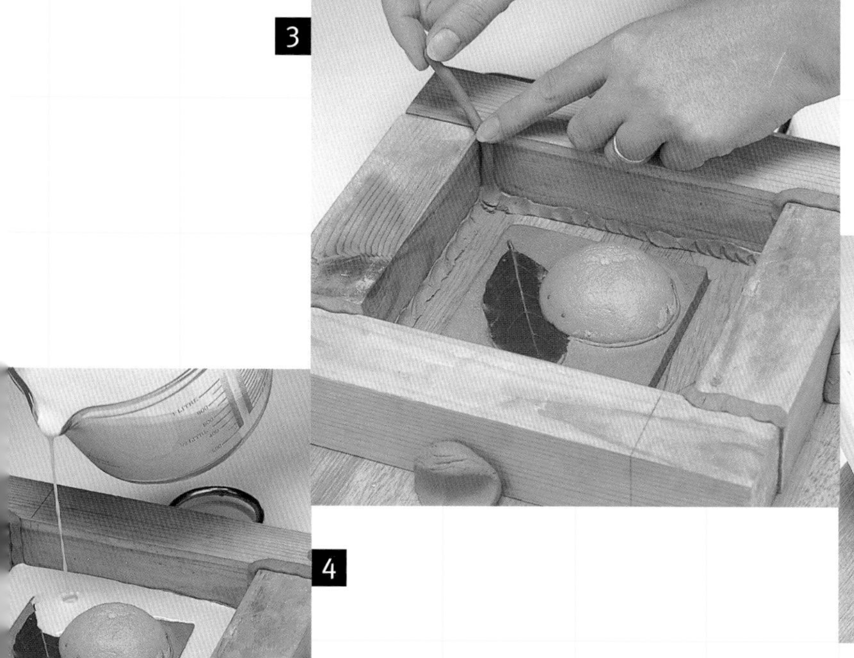

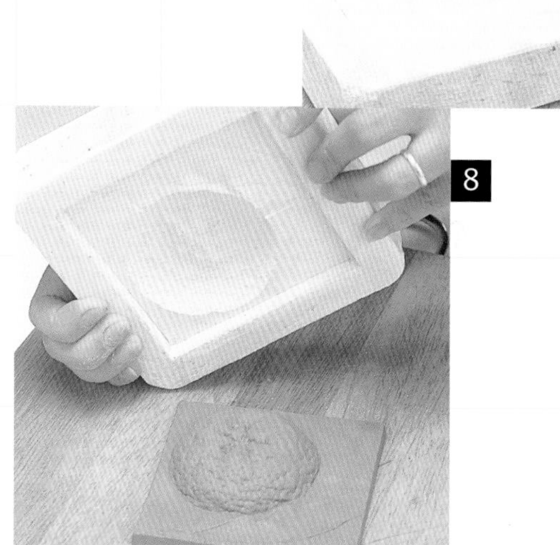

around its base to seal it to the board. Arrange half the orange and its leaf on the tile and tack it in position with the dressmaker's pins. Again, use the moistened paintbrush to seal the edges of the fruit and leaf.

3 Use the lengths of wood to make a frame around the tile, leaving a 1-inch (25mm) gap between the clay and the frame and making sure the frame height is at least 1 inch (25mm) above the fruit. Seal the corners of the frame and the junctions between the frame and the baseboard with thin coils of clay.

4 Pour the prepared plaster (see page 65) slowly into the frame, filling it to the top.

5 Gently pat the plaster surface with the palm of your hand. This disperses any air bubbles and pushes the plaster into the detail of the fruit.

6 Leave the mold to dry until it becomes whiter and warm to the touch. Ease away the wooden frame and gently smooth any rough edges on the outside of the mold with a surform. Turn the mold over and remove the clay tile model. Then, using tweezers, carefully but firmly lift out the fruit and leaf.

7 Take a ball of clay and press it into the mold. Using your thumbs, push it firmly into all corners. Pull a straight strip of wood across the clay surface to remove any excess.

8 When the clay has shrunk away from the edges of the mold, turn the mold upside down and tap the base to loosen the tile.

9 After biscuit firing, paint on the details using transparent glaze tinted with glaze stains (the proportion by weight of the stain should be no more than 4-10% of the glaze). Finally, spray on a coating of colorless glaze.

Firing

Follow the programs for (1) low biscuit firing, (2) earthenware glaze firing (see page 58).

SAFETY NOTES

Handle glazes and stains in accordance with the health and safety precautions recommended by the supplier or manufacturer. When spraying, always wear a protective face mask and either work outdoors or in a spray booth.

Slip casting is a way of creating large tiles without the minor imperfections of those made from plastic clay by extrusion, rolling, or pressing. The casting forms a tile within a plaster mold by the passive accretion of clay from a liquid slip.

Slip-cast Tiles

Equipment & Materials

- *Ammonite shell (or a plaster model of one)*
- *Smooth potter's clay*
- *4 mold boards (see page 51)*
- *Large sheet of glass*
- *Carpenter's square*
- *110 pounds (50kg) pottery plaster (see page 65)*
- *Drill with standard bit*
- *Jigsaw or keyhole saw*
- *Rolling pin*
- *Piece of plastic*
- *Potter's kidney*
- *Soft soap*
- *8 plastic mold natches*
- *Strip of wood*
- *4 plastic 35mm film cases*
- *Surform*
- *Wooden board*
- *2 webbing straps*
- *Large plastic container*
- *Several bricks or concrete blocks*
- *4½ gallons (20 liters) red earthenware slip (20% grog)*
- *Clay knife or spatula*
- *Aerosol spraygun*
- *Protective face mask*
- *Basalt black slip or linseed oil*

The tile leaves the cast with an inherent stillness which is in marked contrast to the more energetic and irregular surface of a hand-formed tile. This quality can be accentuated if the model used to make the mold has a perfect machined or molded surface. Handmade decorative tiles can thus be created with a tranquil and pristine finish in keeping with a modern interior.

High-relief wall panels can be cast as hollow forms. Complex natural forms, such as the fossilized ammonite in this project, can be used as the original model, and having created a single mold, you can cast up to a 100 replicas. With only a small amount of modeling, the fossil can be made to stand out in sharp relief as if it had just been revealed on a freshly cloven rock.

As the weight of the two-piece mold used here (especially when it is full of slip) is considerable, this project should only be attempted if you have someone who can help you lift the mold during its construction and use.

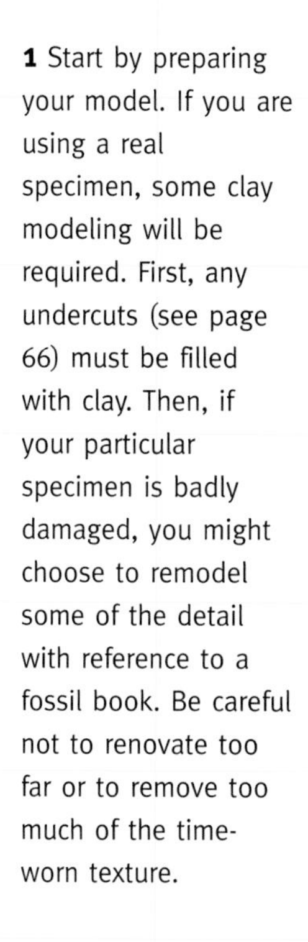

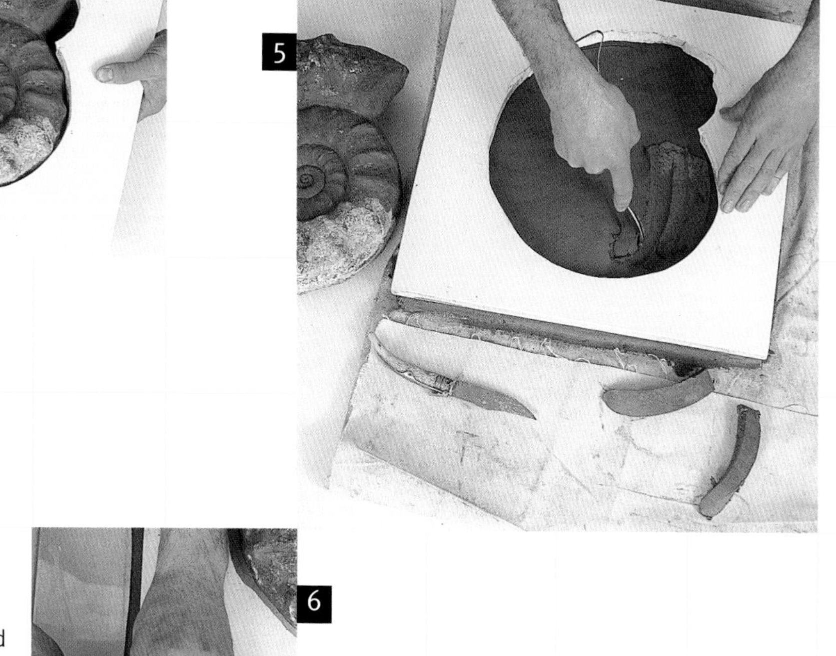

1 Start by preparing your model. If you are using a real specimen, some clay modeling will be required. First, any undercuts (see page 66) must be filled with clay. Then, if your particular specimen is badly damaged, you might choose to remodel some of the detail with reference to a fossil book. Be careful not to renovate too far or to remove too much of the time-worn texture.

2 Arrange the set of mold boards on the sheet of glass to form an 18 × 18-inch (450 × 450mm) square frame. Check all the corners with a square and make sure the two diagonals are equal. Secure the boards with some stiff plastic clay pushed up against the outside of each corner. Apply a thin coil of clay to the seams inside the frame, between adjacent boards and the boards and the glass, then scrape this clay back to leave only a thin fillet along each seam. Pour the mixed plaster into the frame to cast a square slab of plaster ¾-inch (15mm) thick (see page 65). This will form the upper surface of your model.

3 When the plaster has set, remove the mold boards and turn the slab over. On the smooth surface that was cast against the glass, mark out the profile of the ammonite. The area within this has to be cut away without damaging the surrounding slab. First, drill a hole within the central area large enough to insert either a jigsaw blade or a keyhole saw. Then, supporting the slab on a bench, carefully saw around the profile.

4 Carefully place the template over the top of the ammonite, checking that it fits over the widest profile with no more than a ½-inch (10mm) gap anywhere.

5 Take a block of clay and, with a rolling pin, roll out a slab 1½ inches (40mm) thick and at least ¾ inch (20mm) wider than the plaster template. Gently bed the plaster template squarely onto this slab, with the face that was cast against the glass uppermost. Scoop out enough clay from within the cutout area of the template to be able to set the ammonite into position with its widest point level with the plaster surface. Place both ammonite and clay slab on a large sheet of glass resting on a strong work table. Make sure the table is level and the template is bedded evenly.

6 Working on the overlapping margin of clay around the outside of the template, create outward sloping faces on the edges of the panel. Begin by pushing some of the clay up against the side of the plaster template and then, using a piece of plastic, gradually hone down the whole edge. Check that the base of the slab is square, then finally smooth the clay with the kidney.

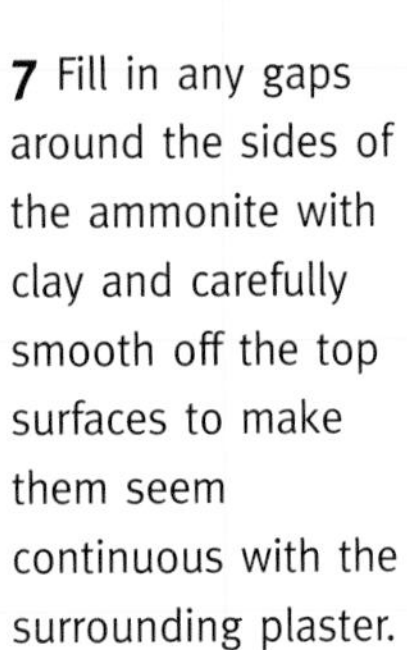

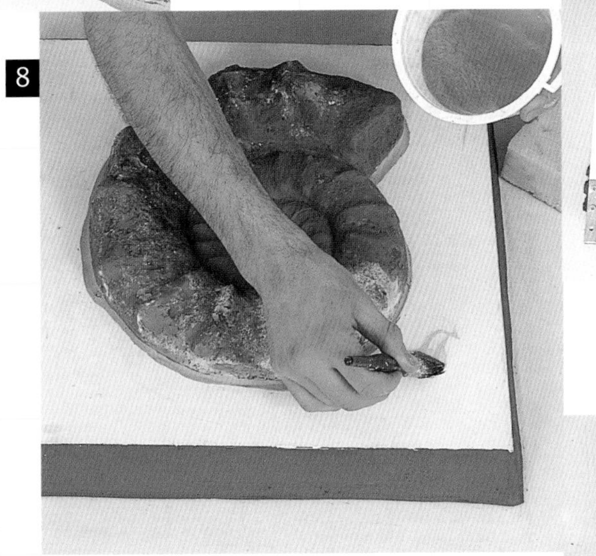

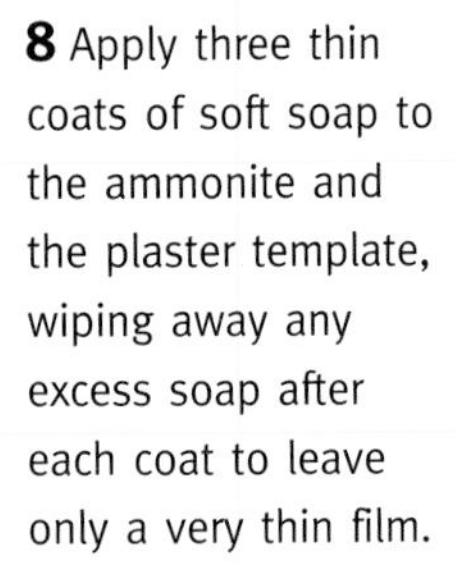

7 Fill in any gaps around the sides of the ammonite with clay and carefully smooth off the top surfaces to make them seem continuous with the surrounding plaster.

8 Apply three thin coats of soft soap to the ammonite and the plaster template, wiping away any excess soap after each coat to leave only a very thin film.

9 Place the four mold boards around the finished panel, leaving a 1½-inch (40mm) gap on all sides. Secure each board by pushing lumps of plastic clay firmly against its outside face. Check that the boards are square to the model and then "clay up" the inside seams.

10 Position four plastic mold natches around the panel, halfway between its clay edge and the mold boards, first filling each natch to excess with clay. When they are pressed down, the clay sticks them to the glass. These act as plugs and sockets to help align and hold sections of the mold together.

11 Pour the prepared plaster into the mold, allowing for a depth of 1½ inches (40mm) above the highest point of the ammonite. Even with a large mixing bucket, this mold will need several pours. With the first pour, run the plaster down the side of the mold boards to avoid dislodging the mold natches or eroding any of the clay modeling on the ammonite. With the second pour, gently cover the entire surface of the model with a thin coat of plaster. This ensures that the junctions between subsequent pours are not visible on the inside of the mold. Each pour will harden before you are ready to pour again, so slightly roughen its surface to help bond with the next.

12 When the plaster has hardened, remove the mold boards and with the help of a companion, turn the mold over to expose the clay back of the panel. If the model falls out, firmly pack the whole of the mold with clay. **Do not replace the model as you are likely to damage the delicate inside surface of the mold.** Drag a wooden strip across the clay back of the panel to check that it is still even and level with the plaster walls of the mold. Now make the formers for the four pour-holes: take

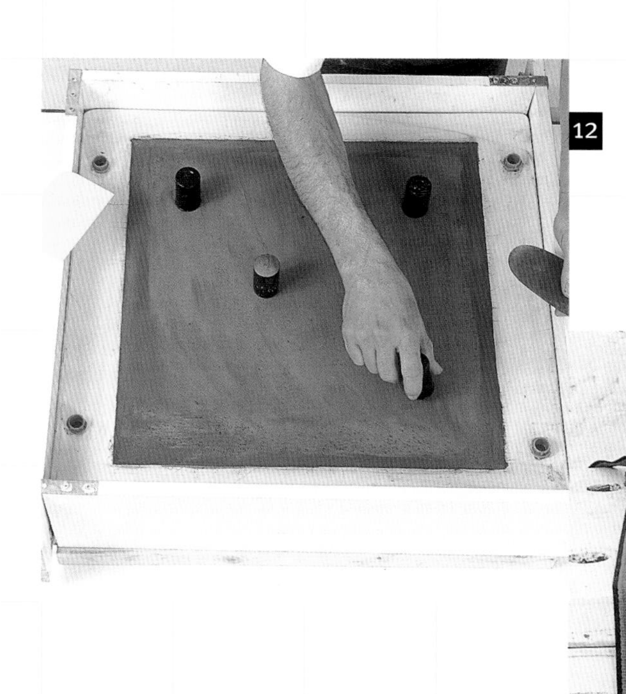

12

13

14

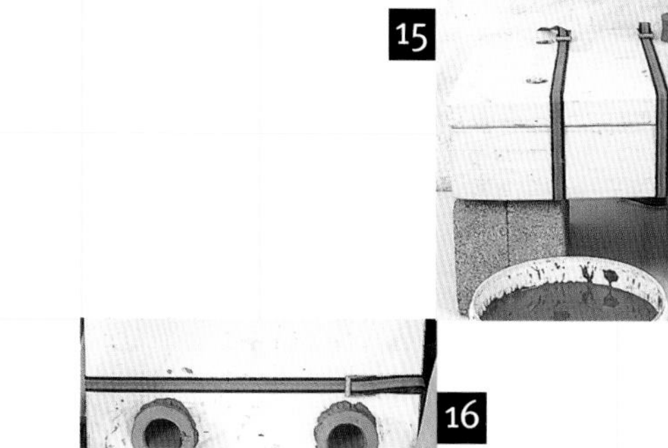

15

16

four 35mm plastic film cases and fill them with enough clay to leave a small cone projecting at the end. Position each case on the panel, 3 inches (80mm) in on the diagonal from each corner, and press down firmly, squashing the clay cones. Reset the mold boards, ready to make the back-plate of the mold. Soft-soap the exposed areas of the plaster and the boards, and insert four more natches into those already set in the mold. Pour the plaster, again allowing for a depth of 1½ inches (40mm).

13 When the back-plate has hardened, remove the boards and smooth off the outside edges of the plaster mold with a surform. Remove the back-plate, place a wooden board across the open mold, and turn it over. After 24 hours the model should have released itself. Remove it from the mold and carefully clean out any loose pieces of plaster or dried clay with a soft brush.

14 Strap the two sections of the mold firmly together with strong webbing. Find a large strong plastic container, and on one side of it build up a stack of bricks or concrete blocks level with its top. Place the mold across the gap between the container and the blocks. Shape clay collars around two of the pour-holes.

15 Start to fill the mold with the grogged casting slip, alternating between the two holes. Continue filling until the slip appears at the top of the non-collared holes. Leave the slip in the mold for about an hour. The exact time will vary with the type of slip and the porosity of your mold. You need to cast a clay wall around ½-inch (5mm) thick. Monitor the level of the slip in the pour-holes throughout the casting and keep it topped up.

16 Again, seek help to empty the mold as it is now at its heaviest. Place a broad, sturdy piece of wood across the plastic container, then push the mold upright so that it rests with one end on the wood over the container. Leave it in this position until the slip has stopped flowing through the holes.

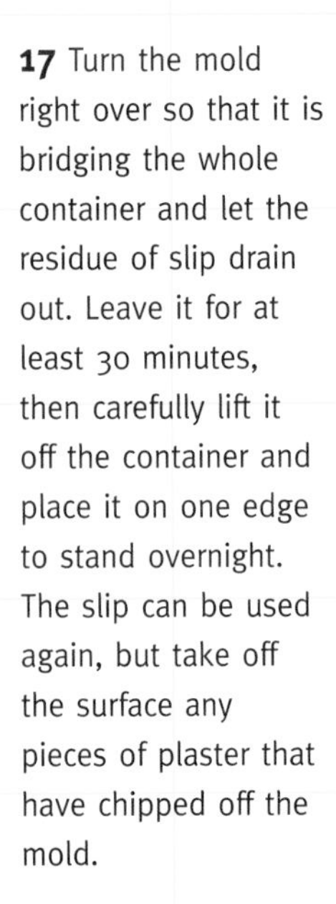

17 18 19

17 Turn the mold right over so that it is bridging the whole container and let the residue of slip drain out. Leave it for at least 30 minutes, then carefully lift it off the container and place it on one edge to stand overnight. The slip can be used again, but take off the surface any pieces of plaster that have chipped off the mold.

18 Using a clay knife or a narrow spatula, carefully cut away the spare clay that has deposited inside the pour-holes.

19 With the mold still standing on its edge, undo the webbing straps and take off the back plate. Gently press the back of the panel to see how stiff the clay wall has become. When it no longer flexes easily and all the wet sheen has gone, it is ready to remove from the mold. The top surface of the panel carries the ammonite relief and is comparatively heavy. To stop it from collapsing inward when the panel is laid on its back, you must insert five props of stiff plastic clay. Push one into each of the pour-holes as far as it will go and cut a fifth hole directly below the center of the ammonite for the last. Trim off the ends flush with the back of the panel.

20

20 Place a board across the back of the panel and turn the board and mold over onto a flat surface. Lift the mold off the board to reveal the cast panel underneath. This can be colored by spraying a thin layer of vitrifying slip (e.g. basalt black slip) over the surface or left the natural terracotta and simply sealed after firing with a linseed oil finish. This preserves the fine detail of the ammonite and the raw natural feel of the panel.

Firing

Follow the program for high biscuit firing (see page 58).

SAFETY NOTE

If you spray a slip on your finished tile, always wear a protective mask and work in a well-ventilated area, preferably outdoors.

Decorating Biscuit-Fired Tiles

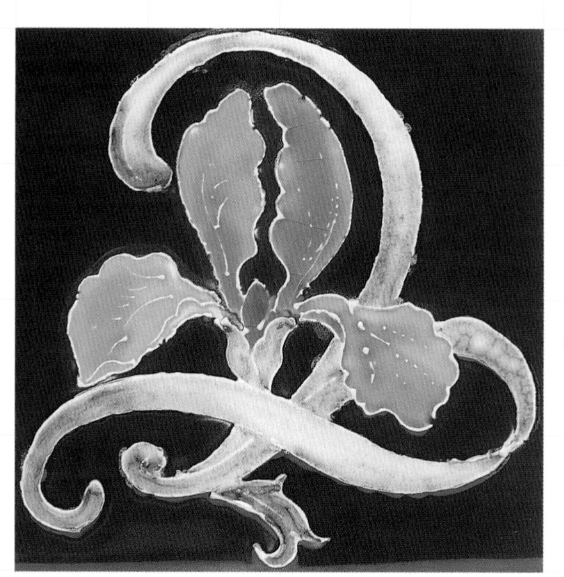

Biscuit tiles, which have been prefired but not glazed, can be purchased for use by the home ceramicist. They have a smooth, flat, absorbent surface that is ideal as a base for experimenting with designs and decorations. The projects in this section use water based decorating media that, in a glaze firing, provide a permanent finish. Using rich colors and flowing lines you can reproduce, or adapt according to your artistic inclinations, the styles of the Orient, the Middle East, and the distinctive Art Nouveau movement of the early 20th century.

The techniques demonstrated in the following projects (pages 101–19) can be used to decorate, with a permanent ceramic finish, pre-fired tiles purchased from a tile retailer or craft supplier.

Which Tile?

Unglazed Biscuit Tiles

No definitive rules can be given on the suitability of any commercial tile to a particular decorative technique, but the general guidelines given below will help you select which pre-fired base tiles are worth testing.

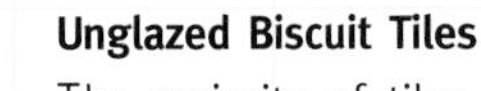

Handmade Biscuit Tiles

Preglazed Tiles

Unglazed Biscuit Tiles

The majority of tiles on sale are glazed, but a limited selection of unglazed biscuit tiles is also available. These are generally for use on counters or floors and should be purchased in an untreated state (i.e. without any surface oiling or waxing). Some craft suppliers also stock unglazed wall tiles specifically for the home ceramicist.

Biscuit wall tiles are usually dust-pressed and made from a fine-grained white earthenware clay. They are highly absorbent and will remain flat and even in an earthenware glaze firing.

Also readily available are unglazed terracotta floor tiles. These are made on an industrial scale by pressing or extruding plastic clay and are generally up to ½ inch (15mm) thick and 8-12 inches (200-300mm) square. They are made of a red clay body which appears fine and smooth on the surface. They are usually still absorbent and can be readily glazed.

Handmade Biscuit Tiles

Handmade biscuit tiles are available from specialized retailers and local potteries. They can vary greatly, but are generally coarser in their fabric and finish than any industrial product. A variety of sizes is available ranging from moderately thin wall and counter tiles, through to large heavy floor tiles. The clays can be buff or red earthenware, but most are low firing.

The amount of nonabsorbent grog that is exposed on the surfaces of such tiles can make them difficult to glaze. The rougher the surface texture, the more speckled and pinholed the fired glaze will be.

Reclaimed Tiles

It is possible to obtain and glaze old tiles from an architectural reclamation center. This would allow you to set a decorative panel within an antique floor. As these are likely to be either contaminated with organic material or saturated with water, they must first be slowly refired to a biscuit temperature before any glaze is applied.

Preglazed Tiles

Four of the following projects deal specifically with on-glaze decoration where enamel pigments are fired onto an already glazed tile. These pigments are formulated with a combination of colorant and flux (see page 62) that allows them to fuse to a glaze at temperatures as low as 1,382°F (750°C). This is well below the melting point of most industrial and studio pottery glazes.

Care should be taken when refiring handmade tiles which have either a crackle glaze or a richly colored transparent glaze. These finishes can be based on low-firing alkaline fluxes which might start to remelt even at low enameling

Designs drawn or painted in wax beneath richly colored stoneware glazes have the subtle and ancient quality of oriental ceramics. Any type of biscuit tile can be used.

Wax-resist Panel Design

Equipment & Materials

Photocopy of design
Ruler
Pencil
Biscuit-fired tiles
Tracing paper
Scissors or craft knife
Carbon paper
Small saucepan
Paraffin wax
Beeswax
Tjanting pen
Oil-spot or Tenmoku glazes
Wide bowl
Alternatives:
Wax emulsion
Paintbrush

In the technique of wax-resist, a wax film covering the outlines of a design is used to repel the application of a water-based glaze. During firing, the wax burns off to expose the contrasting color of the underlying clay. Variations in the depth of the glaze, created as it runs off the resist, add texture and vigor to the surface. Traditionally, the wax (as used for candle making) is applied as a hot liquid, solidifying as soon as it touches the cold absorbent surface of the ceramic. It is poured from a small ladle called a "Tjanting pen," which is designed for drawing batik patterns on fabric. A modern alternative is to use a cold water-based wax emulsion. On a biscuit clay surface, the water is absorbed from the emulsion, leaving the deposit of wax. In its most basic form, a resist can be drawn directly on a tile with a wax crayon.

This project uses a design composed of strong flowing lines to cover a panel of six tiles.

1

1 Choose a design and adapt its size on a photocopier to fit the proportions of your tile panel. Mark a set of squares on the photocopy that correspond to the actual tiles, allowing a 1/8-inch (3mm) gap between each square for the grout that will eventually separate them. Place a sheet of tracing paper over the photocopy and pencil in the sections of the design that will occupy each tile.

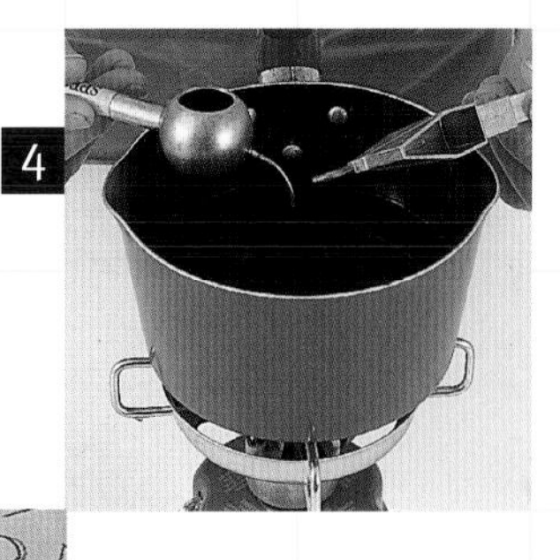

2 Cut out the individual squares and transfer the images to the tiles with a sheet of carbon paper.

3 In a small saucepan over a low heat, melt the paraffin wax and beeswax together in a ratio by volume of 2 to 1. Do not allow the wax to boil.

4 Dip the Tjanting pen into the hot wax and leave it there to heat up. The wax fills the small metal bowl of the pen and will flow out through the narrow metal spout, that acts as the nib. The larger the spout, the more bold the drawing will be. When you are ready to start drawing, turn off the heat. The wax will remain warm and molten for a considerable time.

5 Lift the pen out of the wax and wipe the outer surface of the bulb with a paper towel to remove any drips. Start trailing the wax around the design in smooth continuous strokes. The wax cools in the spout and is quite viscous by the time it reaches the end. With narrower nibs it will flow easily only when the tip is kept in contact with the surface of the tile.

6 If you are using wax emulsion, this is generally applied with a paintbrush and is easily thinned down with water. It dries rapidly on the brush, which will require periodic soaking in warm soapy water to keep it supple and fresh.

7 Pour the glaze into a wide bowl. As hot wax will cool within seconds, the tile can be dipped immediately. Wax emulsion, especially if applied thickly, must be left to dry. If it is still wet, it may run when the tile is moved and will also contaminate the glaze.

8 Dip the tiles with the decorated face uppermost. This allows you to control the run-off of the glaze from the wax by angling the tiles as you lift them out. Some trials will be required to get the consistency of the glaze right (see page 61). It must be applied as thickly as possible, yet still able to clear from the waxed areas. Leave each tile for several minutes after dipping to determine success, as a thick glaze will take time to peel slowly back from the wax.

Firing

Follow the program for stoneware glaze firing (see page 58).

SAFETY NOTE

When handling glazes, always wear protective gloves and take appropriate health and safety precautions as advised by the supplier or manufacturer.

Card stencils are simple and cheap to make, and you can use them to decorate a wall or panel of tiles with a flowing geometric border in the classic Greek or Arabic style.

Card Stencils: Greek Key Motif

Equipment & Materials

- *Photocopy of motif, or graph paper for drawing*
- *Carbon paper*
- *Large sheet flexible poster board*
- *Craft knife*
- *Biscuit-fired tiles*
- *Tin glaze*
- *Wooden board*
- *Ruler*
- *Pencil*
- *Masking tape*
- *Protective gloves*
- *Blue majolica pigment*
- *Shallow bowl*
- *Small stencil sponges*
- *Alternatives:*
- *Drill with standard bit*
- *2 short plastic or metal rods*
- *Blotting paper*
- *Protective face mask*
- *Airbrush or stiff toothbrush*

The stencil acts as a mask over which majolica pigments can be sponged, sprayed, or flicked onto the unfired glaze. When removed, it reveals a bold negative design on a strongly colored background. The three methods by which the pigment can be applied produce distinctive textures which soften as the glaze melts to form rich and interesting surfaces. Finer details can be added by carving back the bold blocks of color to reveal the lighter glaze beneath.

1 Photocopy a motif from a source book to the required size or design your own by drawing it on graph paper (see page 36 for instructions on drawing the Greek key pattern). Repeat the design several times to see the effect of the finished border and how the motif might be modified to turn a corner. Trace the design with carbon paper on poster board and draw a tab projecting out

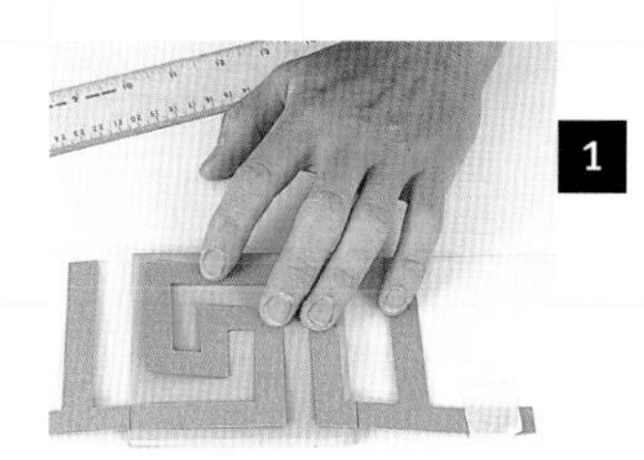

1

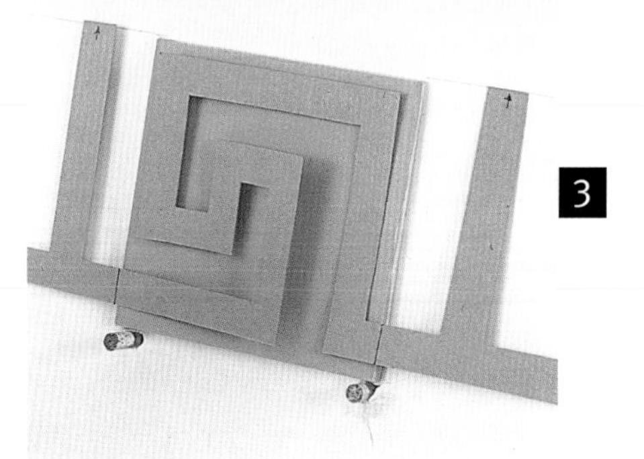

3

4

5

from each side of the design (these will help to align the stencils in place while the pigment is applied). Cut out the motif and the tabs with a sharp craft knife.

Coat a biscuit-fired tile with tin glaze (see page 62) and, when dry, place it on a wooden board with one edge against a pencil line. Place the stencil over the tile, with the tabs aligned on the same line. Fasten the tabs carefully with masking tape.

2 Wearing protective gloves, mix the majolica pigment with water in a shallow bowl. Soak up some of the wet pigment with a stencil sponge, squeezing out any excess. Dab the sponge over the stencil and tile to give an even layer of color. Always experiment before attempting a run of tiles.

3 Flicking or spraying techniques require the tile to be held in a near-vertical position. Drill two holes in the wooden board just below the baseline and insert short pieces of plastic or metal rod to act as supporting pegs. Cover the board with a sheet of blotting paper so any excess pigment can be easily cleaned up and discarded.

4 To spray, fill the airbrush jar with the mixed pigment, making sure there are no hard lumps to block the nozzle. The pigment will constantly settle to the bottom of the jar, so shake it before spraying to keep the color density even. Hold the airbrush approximately 8 inches (20cm) away from the board and adjust the nozzle to achieve a good density of pigment without delivering large droplets. Using steady, sweeping movements, spray the tile until it is evenly coated. Make sure you do not overdo the edge of the tile when you change direction at the end of each stroke. Several coats may be necessary.

5 Flicking the pigment onto the tile produces a more textured surface than either of the previous two techniques. Dip a stiff toothbrush into the majolica mix and shake off any excess. Holding the brush about 6 inches (15cm) away from the tile, draw your thumb toward you over the bristles, flicking the pigment back against the tile surface.

6 After applying the pigment by any of the above techniques, let the surface dry and then carefully lift off the stencil to prevent any beads of pigment from running onto the tile. Place the stencil on a clean piece of newspaper and wipe the top surface clean with a dry sponge.

Firing

Follow the program for earthenware glaze firing (see page 58).

SAFETY NOTE

- When handling majolica pigments, always wear protective gloves and refer to the manufacturer's guidelines on health and safety.
- Spraying or flicking techniques produce minute air-borne particles of pigment and should only be attempted outdoors or in a spray booth. Protective gloves and a face mask must both be worn.

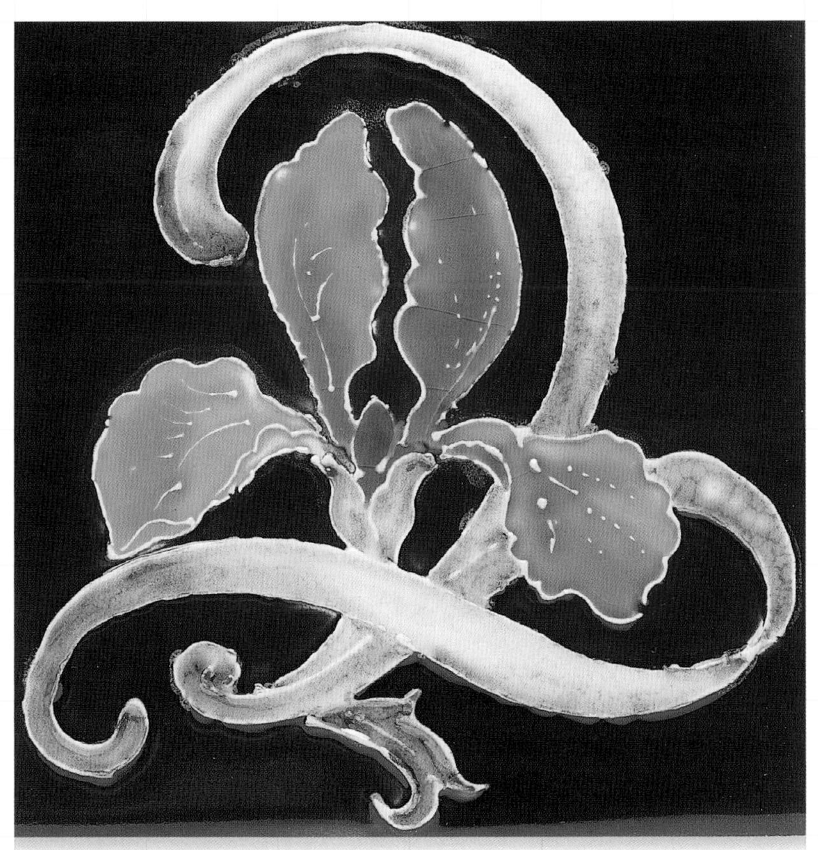

The extravagant flowing forms of the Art Nouveau designers were reproduced in all elements of interior design and became a popular theme for the tile industry. Using the tube-lining technique, you can recreate this richly decorative look.

Tube-lined Art Nouveau Design

Equipment & Materials

Photocopy of design
Tracing paper
Carbon paper
Biscuit-fired tiles
2 plastic slip trailers (one with 1mm diameter nozzle)
Tube-line mix (see panel)
Paintbrush
Craft knife
4 richly colored transparent earthenware glazes

Original Art Nouveau tiles were produced by pressing moist clay dust into a carved metal mold, which revealed the design as fine raised lines on the tile surface. Richly colored glazes were then flooded onto the tile, with the raised lines acting as boundaries between each color. This style of decoration was developed from a method known as "tube-lining," in which a paste of clay, flux, and quartz is trailed from a bottle or bulb to apply a raised pattern onto a biscuit clay surface. The design pictured in this project was adapted from an early 20th-century brooch to fit within a 6-inch (150mm) square.

1 Either create your own design, or find one in an Art Nouveau source book and photocopy it to the correct size. Trace the design and use carbon paper to transfer it to a biscuit-fired tile with a smooth, even surface.

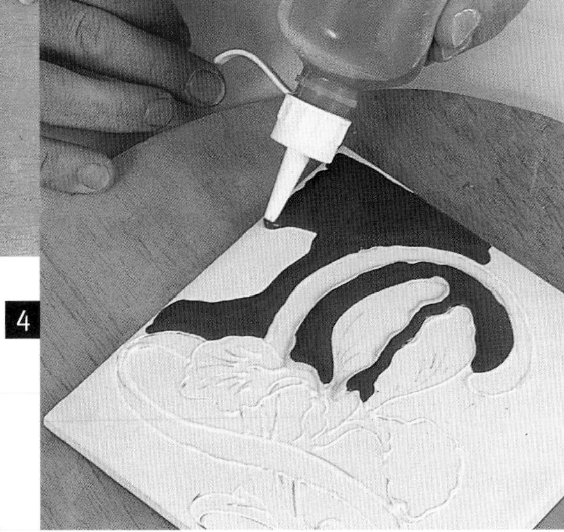

2 Fill a fine-nozzled (1mm diameter) slip trailer with the tube-line mix (see panel). A plastic bottle-type trailer is the easiest to use as the paste can be spooned directly into it. Shake the bottle well before starting to trail, as a watery layer may separate out on top of the mix. If the mix is too thin, it will flow out too freely. To produce a fine continuous line, keep the tip of the nozzle in contact with the tile so the absorbent surface sucks out the mix. Work by drawing the bottle toward you, which keeps the lines smoother and more even. Do not squeeze the bottle unless you are applying dots or filling in large areas.

3 Fine details that are to appear as texture beneath the glazes (rather than as boundaries between them) can be painted on with the tube-line mix, using a brush. Carefully trim or remove any mistakes with a sharp craft knife while the trailing is still damp. If any good lines become dislodged from the surface, dampen them with a fine moistened paintbrush and stick them back in place.

4 When the tube-lining is dry, start to fill in the areas of the design with colored glazes. Flood these onto the surface of the tile using a slip trailer with a wider nozzle than that used for the lining mix. To be able to control their flow make their consistency the same as a brush-on glaze (see page 62). The thinner the glazes, the more care must be taken with their application. Always start with the thinnest and let each one dry thoroughly before applying the next. This technique will require practice, so treat some of your tiles simply as test pieces to get the feel of the materials you are using. Always wear protective gloves when handling glazes.

Firing

Follow the program for earthenware glaze firing (see page 58).

TUBE-LINE MIX

The mix must have three things: clay to bind it together, a flux (see page 60) to fuse the line onto the tile, and a granular material such as quartz to reduce its contraction on firing. The following recipe is suitable if a commercially made mix is not available. The percentages given are by weight.

36% china clay
36% quartz
16% ball clay
4% whiting
4% standard borax frit
4% high alkaline frit

Combine the materials with water, initially making the mix thin so it can be pushed through a very fine strainer. Let it stand for 24 hours, after which time some water will have separated out and can be skimmed off. It then needs to thicken further, so place it in an open container in a warm room to encourage evaporation.

Decorating Pre-glazed Tiles

The next four projects look at different methods for applying permanent colors and images to purchased tiles that have already been glazed. Using vivid enamel pigments, which can be mixed to create an almost endless palette, and simple techniques such as painting, and printing with linoleum blocks or sponge stamps, you can create striking effects. For a truly contemporary look to your tiles you can try printing with a fabric screen, a method widely used in studio ceramics for reproducing highly detailed designs. After firing, all these decorations will remain as a permanent finish.

Despite the large choice of decorated tiles now available, it is not always easy to find the right colors to coordinate with fabrics, wall-paints, or bathroom and kitchen furniture. With enamel washes you can change a plain white tile to whatever color you want.

Colorwashed Tiles

Equipment & Materials

Pre-glazed tiles
Gelatin powder
Saucepan
Soft wide brush
Enamel powdered pigments in 2 colors
Sheet of glass
Palette knife
On-glaze "fat oil" based painting medium
Aniseed turpentine
Large lacquer brush

Enamels offer the largest choice of colors in the ceramic palette and can often be intermixed to give you an infinite variety. Their low firing temperature means that they can be fired onto most tiles without affecting the base glaze (see page 63). However, some Spanish or Mexican crackle glazes have a very low melting point, so a test firing should always be done before purchasing a large quantity of base tiles. There is little or no modification of enamel colors during firing, as occurs with pigments used at higher temperatures, so a color match made when applying the enamel should remain true.

1

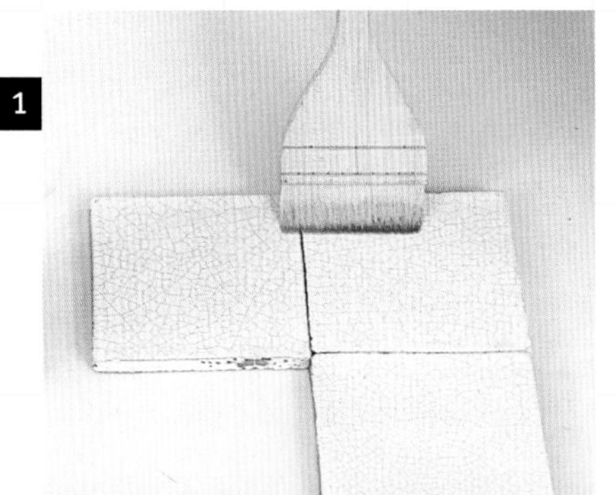

1 To achieve an even application of colorwash on the glassy surface of the glaze, a weak solution of hot gelatin must be applied as a primer. Dissolve the powder in a saucepan of water over moderate heat to make a watery mixture, stirring continually. Do not let it boil. Apply it to the tiles with a soft wide brush. Keep the saucepan warm as you work, as the solution will begin to form a gel as it cools. Any thickened lumps on the tile surface should be immediately smoothed off with the side of your finger. When the coating has cooled, it gives a slight tackiness to the glaze and will seal off any crazing. Fine crazing is not immediately visible, but as soon as a pigment is applied to the surface, it is revealed as minute cracks.

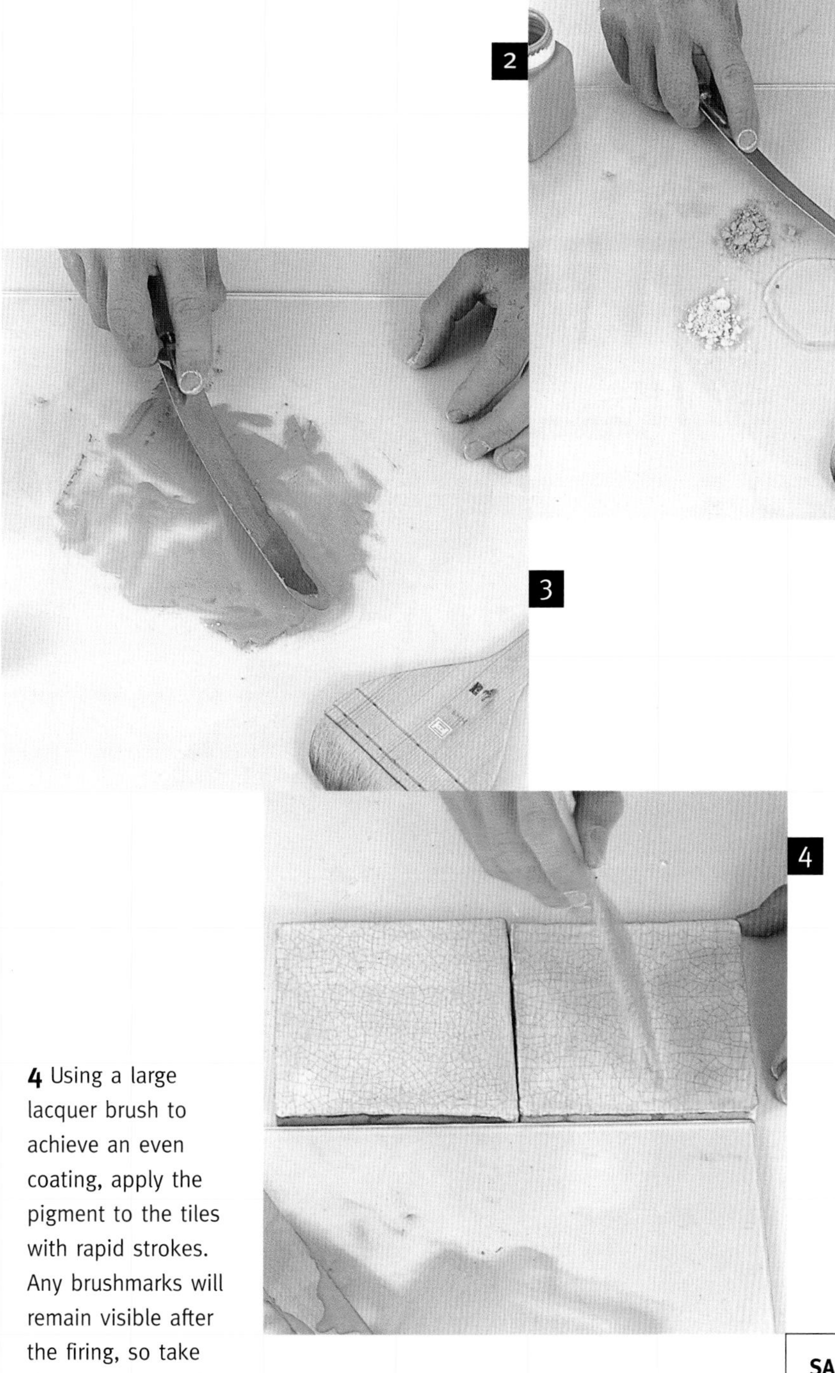

2 Choose two enamel pigments that you believe will mix to make your desired color and place a measure of each on a sheet of clean glass. (The compatibility of the enamels should be checked with the manufacturer before mixing since occasionally the chemical constituents will react to give unpredictable color results.)

3 Using a palette knife, mix the two pigments together with the painting medium, in a ratio by volume of approximately 2 to 1 (powder to medium). Once the desired color has been achieved, thin the pigment with aniseed turpentine to enable it to be painted on as a wash. It is advisable to make a small quantity of pigment for test firing before mixing enough to complete a whole run of tiles.

4 Using a large lacquer brush to achieve an even coating, apply the pigment to the tiles with rapid strokes. Any brushmarks will remain visible after the firing, so take time to produce the right quality of finish.

Firing

Fire the complete set of tiles in a well-ventilated kiln to the enameling temperature, between 1,292 and 1,472°F (700 and 800°C) (see also pages 56–9). The kiln must be fired slowly at first (180°F/100°C per hour) up to a temperature of 932°F (500°C) to allow the medium to burn away, after which firing can speed up.

SAFETY NOTE

As both oil-based painting media and turpentine contain solvents, their mixing and use should be carried out in a well-ventilated area. Take appropriate health and safety precautions as advised by the manufacturer or supplier.

Printing with shaped pieces of sponge is an easy way of producing a simple but bold pattern across a panel of tiles. Once you have tried your hand, you will want to go on to design more complex motifs.

Sponge Design

This project allows you to create a customized tile design with the minimum of equipment. Using enamel pigments will allow you to choose any glazed tile as a base, while the selection of commercially available enamels is large and offers a choice of vibrant colors. The design illustrated here covers a block of nine tiles.

Equipment & Materials

Pre-glazed tiles
Graph paper
Scissors or craft knife
Firm synthetic sponge 2 inches (50mm) thick
Fine felt-tip pen
Gelatin powder
Saucepan
Wide lacquer brush
Enamel pigments
12 × 12-inch (300 × 300mm) sheet of glass
On-glaze "fat oil" based painting medium
Palette knife
Aniseed turpentine
China-marker pencil

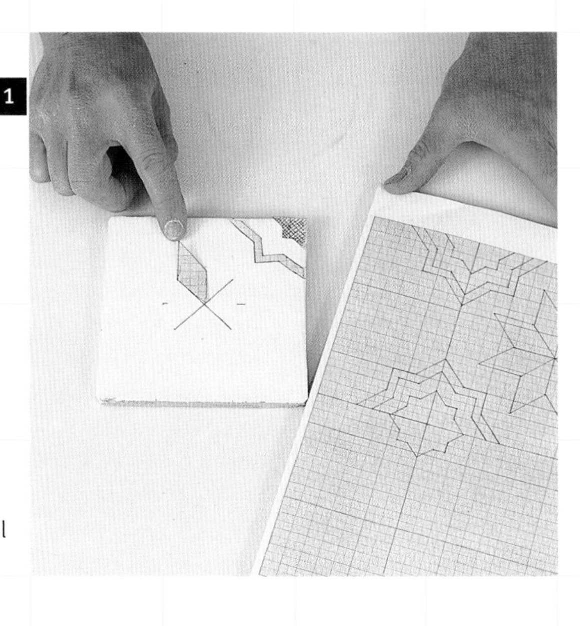

1 Measure the dimensions of your base tile and draw a block of four tiles on a sheet of graph paper. Draw your pattern within the outline of these tiles, keeping the individual elements of the design fairly small since these will be easier to work with. With scissors or a craft knife, carefully cut out each shape you are going to print.

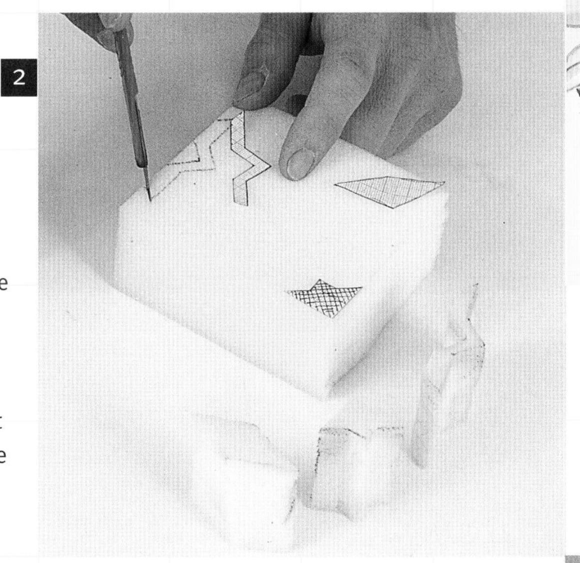

2

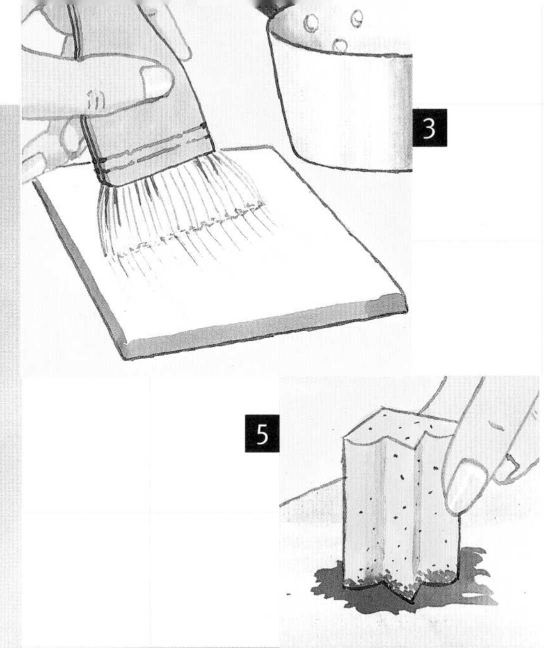

3

5

6

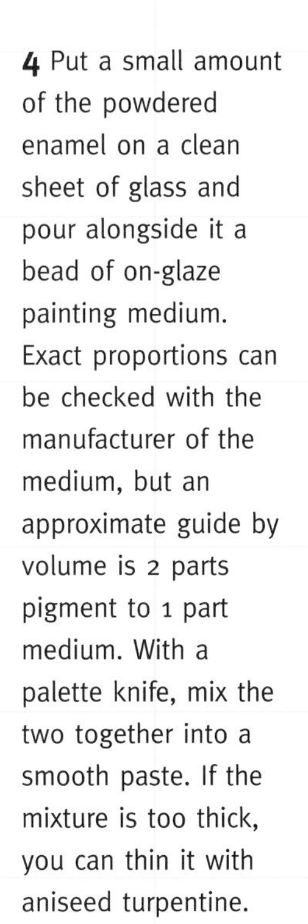

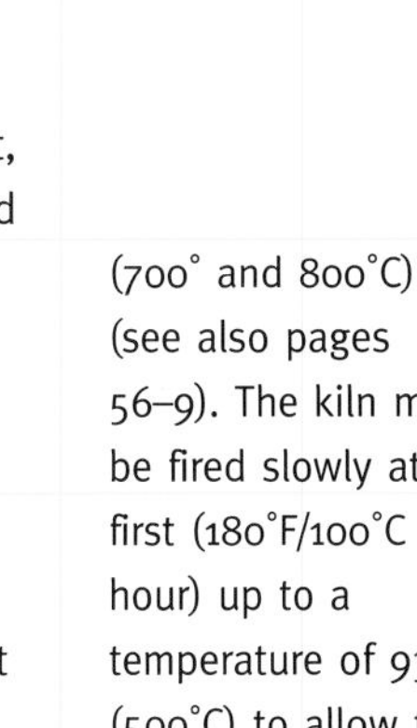

2 Hold each paper shape against one face of the sponge block and with a fine felt-tip pen draw around its edge. If the paper shape moves, try pinning it in place. Using a fine craft knife, carefully cut the sponge from around the shape, making each stamp the full thickness of the sponge. However, only the first ½ inch (10mm) needs to be cut accurately to shape; the remainder is for holding the sponge while you print.

3 Make a weak solution of gelatin to act as a primer on the shiny glazed surface of the tiles. Dissolve the powder with water in a small saucepan over a moderate heat, stirring continually. Do not let it boil, but keep the pan over the heat as you work to prevent the solution from gelling. Using a wide lacquer brush, paint a thin coating over the whole tile surface and leave it to dry. This will help you make a cleaner, crisper print and will burn away during the firing.

4 Put a small amount of the powdered enamel on a clean sheet of glass and pour alongside it a bead of on-glaze painting medium. Exact proportions can be checked with the manufacturer of the medium, but an approximate guide by volume is 2 parts pigment to 1 part medium. With a palette knife, mix the two together into a smooth paste. If the mixture is too thick, you can thin it with aniseed turpentine.

5 Spread the pigment paste in a thin layer on the glass with the palette knife, then ink your stamp by pressing it into the pigment until the color has soaked into the sponge.

6 To help position your stamps, roughly mark out the pattern on the tile with a china marker. To print, gently press the inked stamp against the surface. If your first attempt is too pale, clean the tile with turpentine and start again, making the pigment thicker. Let it dry before firing.

Firing

Fire the complete set of tiles in a well-ventilated kiln to the enameling temperature, between 1,292° and 1,472°F (700° and 800°C) (see also pages 56–9). The kiln must be fired slowly at first (180°F/100°C per hour) up to a temperature of 932°F (500°C) to allow the medium to burn away, after which firing can speed up.

SAFETY NOTE

As both oil-based painting media and turpentine contain solvents, their mixing and use should be carried out in a well-ventilated area. Take appropriate health and safety precautions as advised by the manufacturer or supplier.

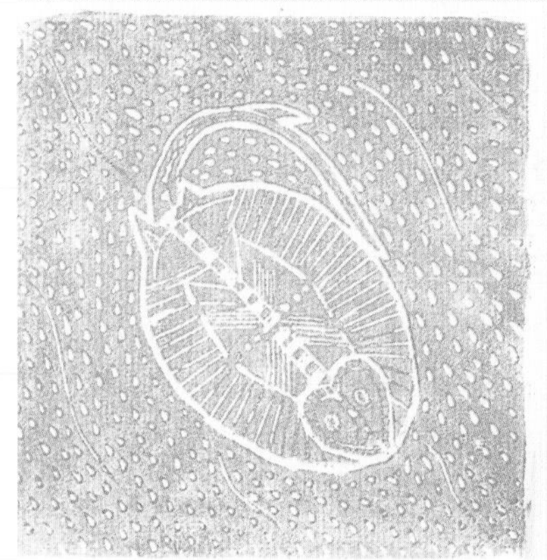

Linoleum blocks allow you to reproduce intricate linear designs set within bold areas of color. Choose a pattern that will be striking as a "negative" print of white lines against a dark background. For example, aboriginal bark-paintings, as illustrated here, are a rich source of imagery.

Linoleum Printing

Equipment & Materials

Linoleum squares, at least ¾ inch (20mm) wider than your tiles
3 small wood blocks
Wooden workboard
Copy of image
Carbon paper
Cutting tools
12 × 12-inch (300 × 300mm) sheet of glass
Enamel powdered pigments
On-glaze "fat oil" based painting medium
Palette knife
Rubber or gelatin print-maker's roller
Sheets of pottery tissue
Pre-glazed tiles
Spoon
Aniseed turpentine

Carving into the surface of a sheet of linoleum is the easiest way to create a printing block. The uncut areas are inked to produce the print, and the cut lines will appear in the color of the base tile. Using on-glaze enamels, you can choose from a large range of intermixable colors.

Linoleum is usually available from artist's supply stores already cut into small squares. When warm, it can be carved easily with purpose-made hand-tools. Most linoleums have an oil-based finish which should be removed with mineral spirits on a rag before you start work.

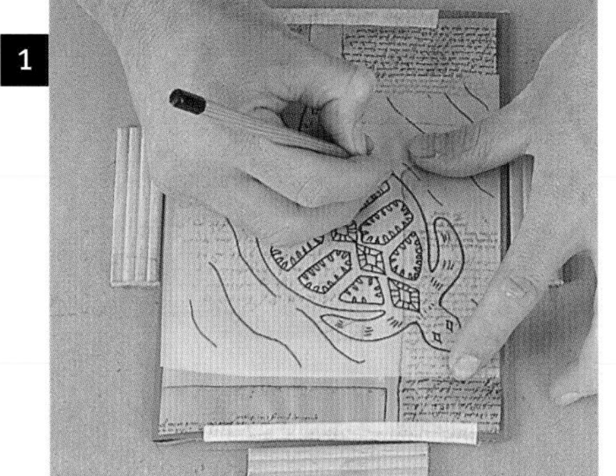

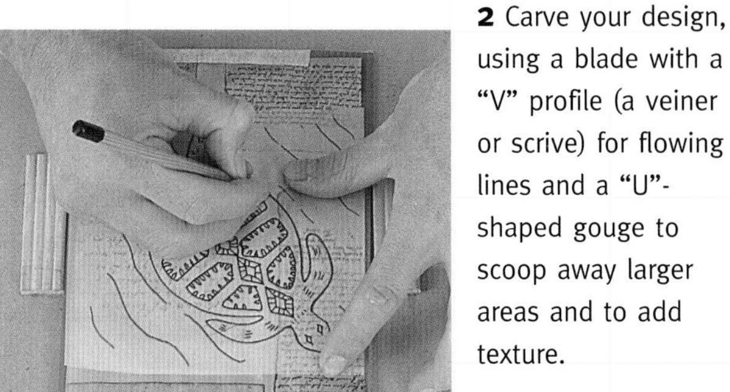

1 Anchor a linoleum square between three small wooden blocks attached to your workboard. Draw or trace your design and transfer it to the linoleum with a sheet of carbon paper. Gently warm the linoleum before cutting (a hairdryer is ideal).

2 Carve your design, using a blade with a "V" profile (a veiner or scrive) for flowing lines and a "U"-shaped gouge to scoop away larger areas and to add texture.

3

4

5

7

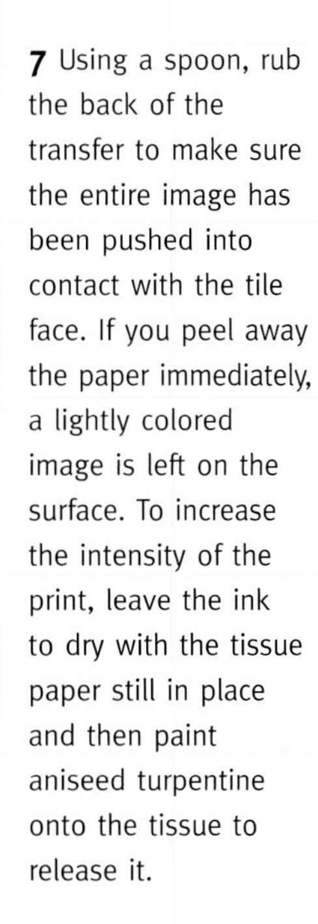

3 On a sheet of glass mix the powdered enamel with the painting medium in a volume ratio of 2 parts pigment to 1 part medium. Use a palette knife to form a smooth paste. Spread the ink into an evenly distributed layer with the roller. If the roller skids, the ink is too thin and must be left to stand while some of the turpentine in the medium evaporates.

4 Again using the roller, transfer the ink onto the block. Push the roller across the glass to pick up an even layer of ink and apply it to the linoleum in a single smooth stroke.

5 Take a piece of pottery tissue and lay it gently on the linoleum. Starting at the center and carefully working out, rub the paper with a fingertip to pick up the ink. By varying the amount of pressure, you can vary the intensity of tone in different areas of the print.

6 Carefully peel the tissue off the linoleum and lay it immediately on the tile. Correct registering at this point is essential since once the transfer is in contact with the tile, it cannot be moved again without smudging the print.

7 Using a spoon, rub the back of the transfer to make sure the entire image has been pushed into contact with the tile face. If you peel away the paper immediately, a lightly colored image is left on the surface. To increase the intensity of the print, leave the ink to dry with the tissue paper still in place and then paint aniseed turpentine onto the tissue to release it.

Firing

Fire the complete set of tiles in a well-ventilated kiln to the enameling temperature, between 1,292° and 1,472°F (700° and 800°C) (see also pages 56–9). The kiln must be fired slowly at first (180°F/100°C per hour) up to a temperature of 932°F (500°C) to allow the medium to burn away, after which firing can speed up.

LINOLEUM PRINTING WITH MAJOLICA

Linoleum printing can also be used with underglaze or majolica pigments on an unfired glazed tile. Printing majolica onto the powdery surface requires a much thinner water-based ink. As the normal surface of linoleum is too smooth to print with this, you must use "flocked" linoleum, which has a thin coating of wool dust glued to its surface. Ready "flocked" linoleum, or the wool dust to do your own "flocking," can be purchased from specialized craft suppliers.

The ink combines a water-based painting medium with the coloring oxides normally used for majolica painting (see page 141). This can be painted onto the linoleum block, or the block can pressed onto an ink-soaked sponge pad. The block is lowered carefully onto the tile and pressed firmly to produce an even print.

SAFETY NOTES

- When handling majolica pigments, always wear protective gloves.
- As both oil-based painting media and turpentine contain solvents, their mixing and use should be carried out in a well-ventilated area.
- Take health and safety precautions as advised by the manufacturers or suppliers of these materials.

Screen printing will allow you to accurately reproduce a photographic or drawn image onto ceramic tiles as many times as you wish. The designs shown in this project reflect the style of the late-19th-century ceramicist William De Morgan.

Screen-printed Designs

Equipment & Materials

Acetate film or tracing film
Liquid opaque (light-opaque ink)
2 18 × 18-inch (450 × 450mm) sheets of glass
Masking tape or photomount adhesive
Thin cardboard
1 roll photosensitive indirect stencil film
4 spring clips
Sheet of black cardboard
Plastic trough
Water hose with fine-spray head
2 wooden framed monofilament screens (200 threads per inch)
Screen preparation paste
Protective gloves
Paper towels
Blotting paper
Rubber printer's roller
Water-based screen filler
Lacquer brush or piece of thin plastic
Gummed paper
Printing table (optional - see page 140)
Sheets of "decal" paper
Ceramic screen printing inks or powdered enamel pigment and "fat oil" based screening medium
Palette knife
12-inch (300mm) long squeegee
Organic-based screen wash
"Covercoat"
"Covercoat" thinner
Ceramic tiles
Fine sponge
Potter's kidney

In essence, screen printing is a method of forcing ink through a finely woven fabric to which various types of stencil are fastened. The stencil acts as a mask to prevent the ink from passing through certain areas of the image. The success and versatility of screen printing lies in modern photosensitive materials that allow you to make complex stencils with relative ease.

Printmakers use various methods of manually cutting masks or painting fillers directly onto a silk screen to produce simple bold stencils. A less direct but more versatile technique is to use a photosensitive emulsion. This can produce a stencil from any drawing or photographic image that has been converted to a light-opaque positive on a transparent or translucent film. The emulsion can be applied directly to the screen as a liquid, but an easier method, as demonstrated in this project, is to use stencil film which consists of a plastic backing already coated with emulsion. During exposure to light (see **Note** below), the emulsion hardens, creating a layer of solid material. If the areas of the design required to print are blocked to light, the emulsion remains soft enough to be washed off. When the plastic backing is peeled away, the end product is a stencil with open spaces through which ink can pass.

For decorating ceramics, the stencil is used to print ink onto "decal" paper which produces "slide-on" transfers that are applied directly to the tile surface.

Note on exposure times:
Exposure time is critical in making a good durable stencil. If the film is under-exposed, the stencil will be weak, and when it is applied to the screen, the edges of the design may blur. It is also possible that the emulsion will be so soft that it will all wash off without preserving the image. Over-exposure can lead to finer areas of the design becoming blocked with hardened emulsion and make the stencil as a whole too hard to adhere to the screen.

Screen Printing Materials
The two essential pieces of equipment are the screen and the "squeegee," a rectangular or "V" shaped rubber blade held in an aluminum or wooden handle. Various sizes are available, and you should choose one that is at least 3 inches (75mm) wider than the image you are printing. The type of fabric chosen for the screen will depend on the fineness of the image being printed, the type of ink being used, and the durability required from the screen. The hole size in the weave that allows the ink to pass through the screen varies with the number of threads per inch, the thickness of each thread, and the style of the weave. The bolder the image, the coarser the mesh you should choose. For printing images with very fine lines or when using pigments that require only a thin deposit of ink, such as metal lusters, a finer mesh is appropriate.

The fabric is stretched over either a metal or wooden frame. Both types can be purchased with the fabric pre-stretched on them. The screen must be 8 inches (200mm) wider than the image you are printing and at least 4 inches (100mm) wider than the squeegee. It is perfectly possible to put two or more stencils on the same screen.

The screen must be held off the printing surface by at least 1/16 inch (2mm), so that as the squeegee passes over the stencil, the screen snaps back up from the surface and prevents the image from smudging. For taking single prints, the screen can simply be held firmly in place on top of two strips of wood with the printing surface raised to the correct height beneath on plywood packing. If you are producing a run of prints, a printing table is essential. The simple wooden construction shown in the drawings can easily be made at home (see page 140).

Inks
To draw or trace the positive image that will produce your stencil, a light-opaque ink must be used. (Normal black drawing ink will invariably prove non-light-opaque.) A thick matte red ink, specifically produced for filling in opaque masks ("stopping out") on negatives is readily available from commercial printers or their suppliers.

For printing, ready-mixed ceramic screen printing inks can be purchased. However, if you want to make your own, any powdered enamel pigment mixed with an oil-based screening medium can be used.

1 To create a positive, draw your image directly onto an acetate film, or trace it on high-quality tracing film, with light-opaque ink.

2 Mount the positive on a clean sheet of glass with either masking tape or photomount adhesive, making sure these materials will not interfere with areas where light will pass through onto the emulsion film. Surround the image with a 3/8-inch (10mm) mat cut from very thin cardboard.

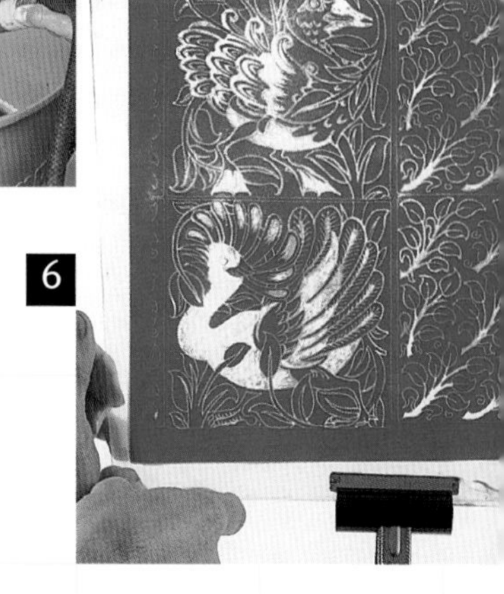

3 The stencil film will generally be supplied as a roll in a sealed tube. Unroll it only in subdued light and cut off the amount you need, immediately returning the remainder to the tube. Cut the film to cover both positive and cardboard mat and lay it on the positive with the shiny plastic backing face down. Place a second piece of glass on top of both to create a sandwich and firmly clamp all together with spring clips.

4 To expose the film, you could use a light source such as a high-wattage metal halide lamp in a purpose-made light box, but the sun will work perfectly well. However, you must work on a day when atmospheric conditions are likely to be constant for several hours. Using a small positive, do several test exposures to establish the optimum exposure time. As a rough guide, a film that needed 2 minutes exposure with a 5KW lamp at 1.2 meters can be exposed successfully after 4½ minutes of bright sunlight. When using the sun, the glass "sandwich" must be backed with a piece of black cardboard to prevent reflected light from the background exposing the film from the reverse side.

5 After exposure, the film must be developed immediately by washing out with a fine spray of cold or lukewarm water. Take the glass "sandwich" out of direct sunlight, delicately remove the film, and clamp it with a spring clip, emulsion side up, to one of the sheets of glass. Stand this in a plastic trough and gently spray water over the surface until all the positive elements of the design are completely clear of emulsion. Wait for one minute and then give it a final rinse down from the top of the film. The resulting stencil must now be applied to the screen before the remaining emulsion hardens further.

6 Before sticking the stencil onto the screen, check that there is no trace of emulsion on any areas of the design that should be open.

7 If you are using a newly purchased screen, it must be pretreated with a preparation paste which roughens the fibers to make them more receptive to the emulsion. Wear protective gloves while applying it. The screen should then be wetted with water before you apply the stencil. Carefully lay the still-wet stencil, emulsion side down, on the bottom of the screen, making sure you place it directly into the correct position. Attempting to slide a stencil across the screen will damage some of the fine edges.

8

8 With the side of your hand, smooth across the plastic backing of the stencil to make sure it is lying flat. Then blot it with a paper towel to absorb any excess moisture.

9

10

9 Invert the screen onto a raised surface (a strong cardboard box is ideal) that allows you to press the stencil into close contact with the screen by gently pushing down on the frame edges.

10 Lay some sheets of cheap blotting paper inside the screen, and with a small rubber roller apply firm but gentle pressure directly down onto the screen surface. If the stencil is correctly exposed, some of the colored emulsion will seep through the mesh and stain the paper. Replace the paper with a dry piece and continue to roll the surface. Repeat until all the moisture is removed from the stencil. If any fluid-containing traces of emulsion are left, they might seal over open areas.

11

11 Leave the stencil to dry at the ambient room temperature for 24 hours. Before removing the backing, fill in the empty areas around the stencil with a water-based screen filler, either painting it on with a lacquer brush or pouring it on and spreading it with a piece of thin plastic.

12 When totally dry, the backing will peel away easily. Any small pieces of stencil that pull away can be painted in with the screen filler. The border around the stencil and any other open spaces on the screen which are not part of the design should also be painted in. Once this filler has dried, the screen is ready to use for printing. Apply a strip of gummed paper around the edge of the screen to act as a reservoir for the ink and to make it easier to clean after printing.

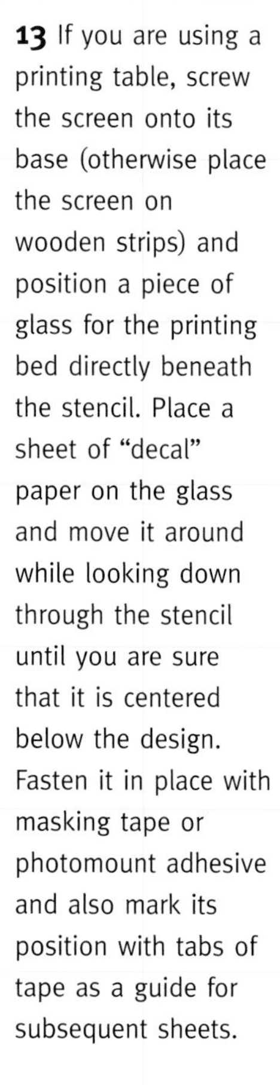

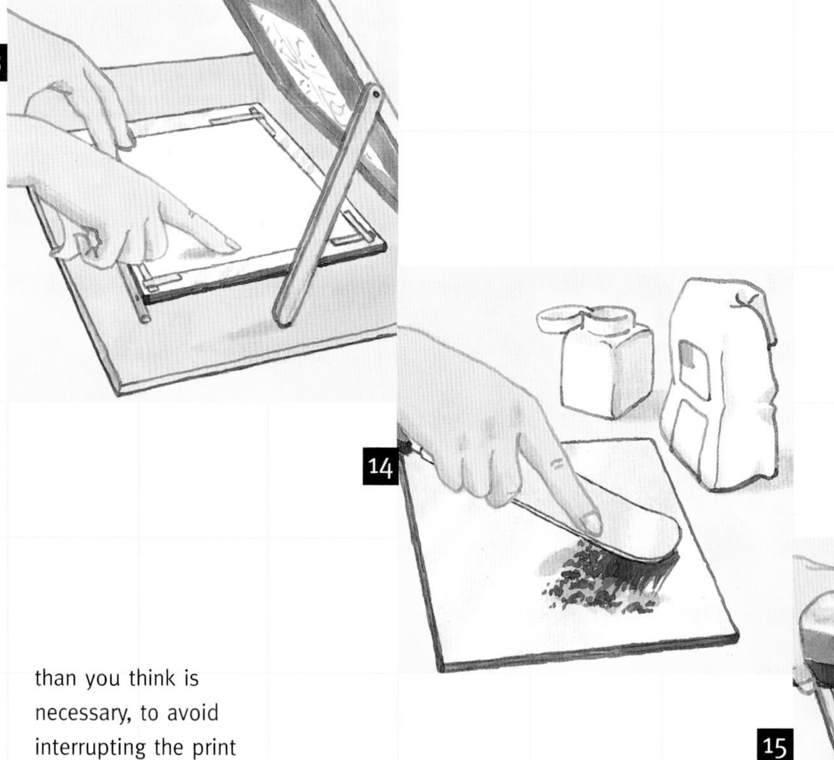

13 If you are using a printing table, screw the screen onto its base (otherwise place the screen on wooden strips) and position a piece of glass for the printing bed directly beneath the stencil. Place a sheet of "decal" paper on the glass and move it around while looking down through the stencil until you are sure that it is centered below the design. Fasten it in place with masking tape or photomount adhesive and also mark its position with tabs of tape as a guide for subsequent sheets.

14 To mix your own ink, blend the powdered enamel pigment with the oil-based screening medium on a sheet of glass in an approximate ratio of 2 parts pigment to 1 part medium (always check with the manufacturers' guidelines). Use a palette knife to mix and make sure there are no coarse or hardened lumps. Surplus ink can be stored in a lidded glass container, so it is worth mixing more than you think is necessary, to avoid interrupting the print run by having to mix a fresh batch.

15 With the screen in its down position, pour an even flow of ink across the border on the near side of the stencil. It will spread out, but its viscous nature will prevent it from flowing onto the stencil.

16 Lifting the screen slightly off the bed, spread the ink across the stencil with a single, gentle stroke of the squeegee. Lower the screen back on the bed, hold the squeegee at 45 degrees to the screen surface and, starting from the line of ink, draw it firmly back across the stencil keeping its top edge horizontal. Repeat this action with each fresh sheet of "decal" paper, flooding the screen on the stroke away from you and printing on the stroke toward you until you have to recharge the screen with more ink. Never leave the screen charged with ink for any time without printing since the ink will dry and harden in the mesh.
When you have completed your print run, use the squeegee to print any remaining ink onto scrap paper, then flood the screen with an organic-based screen wash and push it through to remove any ink residue. (The stencil and screen filler are water-soluble so use only organic solvents to clean them.)

17 Leave the prints to dry on racks or shelves for at least 24 hours. To bind the image on a dry print, a resin film (the "covercoat") must be applied over the top. This can be done with a broad lacquer brush, but on a large run you can print the cover on using a second screen. Simply leave an open mesh area the size of the "decal" sheet and block out the surrounding edge with screen filler. Pour on the covercoat and print as you did with the ink. Clean the screen with a special covercoat thinner immediately after use. Leave the covercoat to dry thoroughly.

18 Several different images can be printed at the same time onto a single sheet of "decal" paper. Cut them out individually, leaving a 3/8-inch (10mm) border around each image. Immerse one at a time into lukewarm water and leave in while the

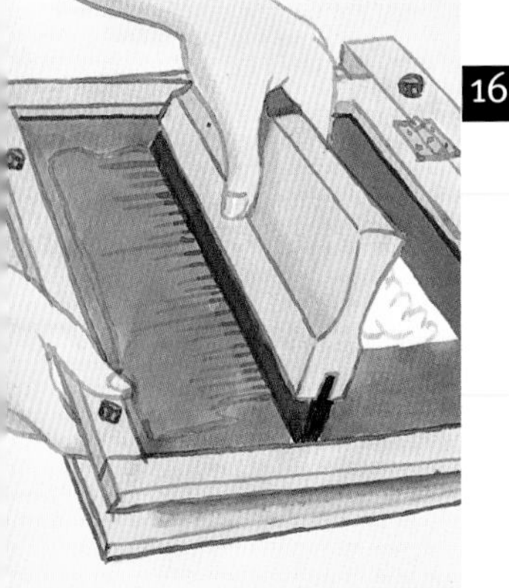

17

backing paper becomes saturated and the gum on the printed surface softens. After an initial curling up as the paper expands slightly, the transfer uncurls again, indicating that it is ready to release from its backing. Remove it from the water and apply a slight pressure to the covercoated surface with your fingertips. The transfer should start to slide away from its backing. Any resistance may be due to a burring of the cut edge. Try applying firmer pressure to the edges only and, if it still resists, leave it to soak longer. When it is free, slide it off the paper directly onto the tile. Once it is correctly positioned, absorb any excess water by gently patting the edges of the transfer with a fine sponge. Use a soft rubber kidney to press it firmly onto the glazed surface and to expel any air or water trapped beneath it.

Firing

Firing of transfers should be taken slowly during the initial phase while the oils in both the screening media and the plastic covercoat burn away. As the kiln heats to 932°F (500°C), the temperature should not rise at more than 180°F (100°C) per hour. The top temperature of the firing should be the normal enameling temperature of 1,382°-1,562°F (750°-850°C). It is possible to print transfers in majolica pigments (see page 141). These should be fired to just below the normal glaze temperature and will result in a softer, more integrated decoration.

SAFETY NOTES

- Oil-based inks, screening media, organic screen wash, and covercoat all liberate solvent fumes. Make sure you work in a well-ventilated area.
- Always wear protective gloves when using a screen preparation paste.
- Take appropriate health and safety precautions as advised by the suppliers or manufacturers of the above products.
- Enamels and transfers should only be fired in a separate and well-ventilated kiln room since noxious fumes will be given off.

PHOTOCOPIED AND HALF-TONE IMAGES

A line drawing or block image can be reproduced onto an acetate sheet with a photocopier, but since it is rarely opaque enough, it either has to be inked in or the stencil film has to be purposely under-exposed. A laser-printed image can be dense enough on white paper if the paper is afterwards coated with liquid paraffin or vegetable oil to render it translucent.

Photographic images can be accurately reproduced, but first their tonal areas must be converted into different densities of fine black dots. This process is called "half toning" and can be done by a local printer. The normal dot density used in a print workshop will be around 150 dots per linear inch. For the screen to accurately reproduce a half-toned image, the mesh count must be four times the dot density. For printing enamels, you must request a half-toned image at no more than 65 dots per linear inch, and if you are printing with underglaze or slip through a coarse mesh, the image should have as few as 30 dots per linear inch. If you specify that you want the half-toned image to be produced as a high-contrast film positive, it will be supplied on a transparent background ready for direct use with stencil film.

Many of the attractive tiles illustrated here can be put to practical use in bathrooms, kitchens, or even on floors. Both handmade and standard biscuit tiles are amongst this selection. Their decorations were achieved with some of the techniques explained in this book.

Gallery of Tiles

▶ LIFE ENHANCING TILE COMPANY

Relief designs
The technique for decorating these encaustic tiles was first used by medieval potters for the flooring of churches. A relief design is pressed into a wet tile, leaving an embossed pattern. The resulting recesses are filled with colored slip and, when dry, the tile is scraped down to reveal the design. The firing process makes very durable tiles for walls or floors.

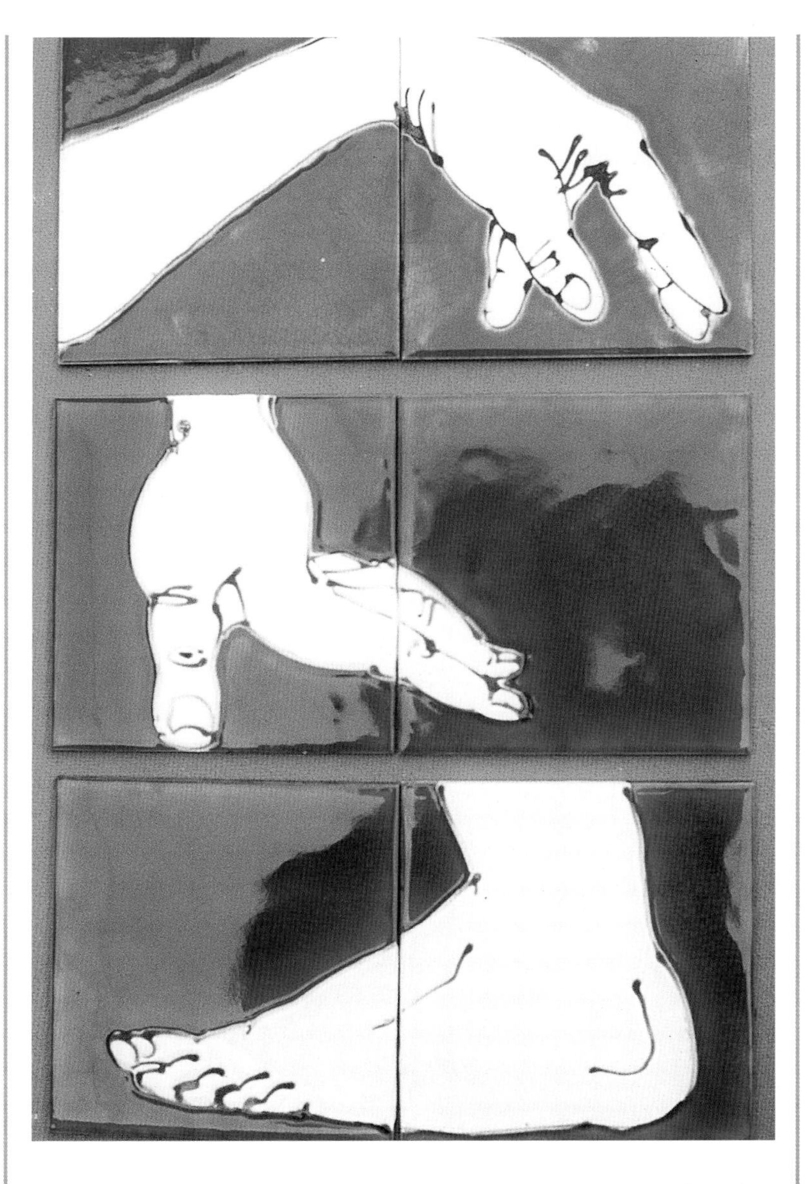

◀ BRONWYN WILLIAMS-ELLIS

Hands and feet
Standard 6 x 6-inch (152 x 152mm) industrial biscuit tiles were used as a base for this series depicting sturdy hands and feet on a deep blue glaze background. The artist has used a traditional Persian wax resist technique, cuerda seca (dry line or dry cord), running the wax directly on to the drawing. Manganese gives color to the lines.

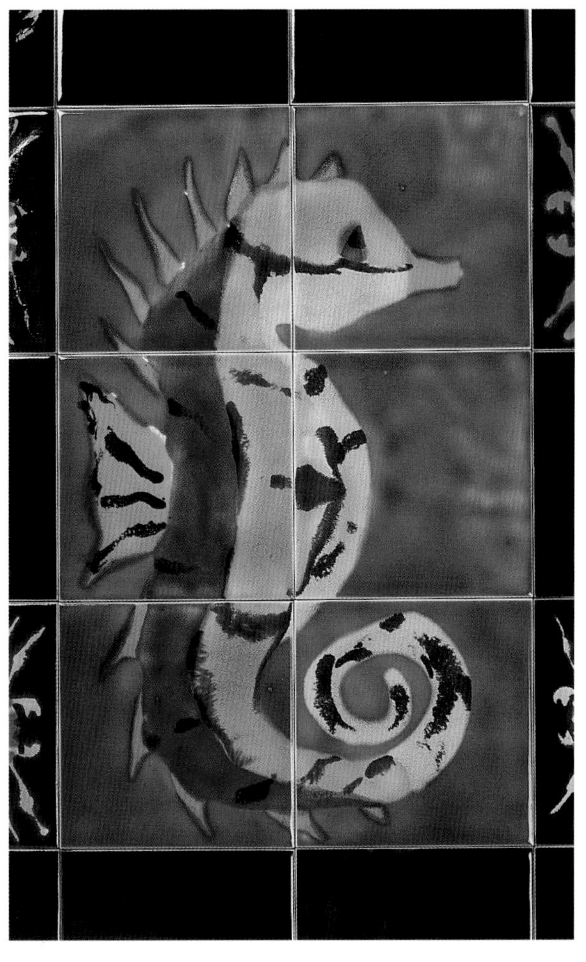

▶ NOREEN JAAFAR

Sea horse
Industrial biscuit tiles, 6 x 6 inches (152 x 152mm), have been used a base for glowing, handpainted colors formulated by the artist from minerals. Their high gloss has the quality of stained glass. The technique is based on a traditional Persian glaze, but the sea horse image is very much the artist's own. The border tiles are decorated with small crabs (just seen in the photograph) to continue the ocean theme.

▲ COUNTRY TILE DESIGN

Sheaf of wheat with corn border
This disciplined relief design is based, like all the artist's work, on a country theme and has an old fashioned feeling with its traditional sheaf of wheat in ripe, golden yellow, its rural tile "frame," and corn border. The tiles are white earthenware, hand-made in the studio. The design is applied in underglaze paints with a transparent glaze on top.

BRONWYN WILLIAMS-ELLIS

Mermaids
Border tiles, cut from 6 x 6-inch (152 x 152mm) standard biscuit tiles, have been used as the basis for this lively group. The artist has created an evocative marine image, which includes a meticulously drawn sea horse, a sea monster, and a merman and mermaid, using various slips and colors to produce a basically gray, white, and yellow scheme.

WORLD'S END TILES

Motif in yellow
A circular motif, embellished with fine ribbon decoration and tiny leaves, covers four tiles. The outline of the design is screen printed on the base and the colors are filled in by hand. This is a traditional design, made in Italy.

MARY ROSE YOUNG

Leaf and cockerel
Images from nature, in a charming and naive style, are depicted on these handpainted tiles. The basic greens and yellows will coordinate well in random arrangements but they could also be used in a separate decorative panel on a wall. Standard biscuit tiles were decorated with underglaze and finished with a transparent glaze.

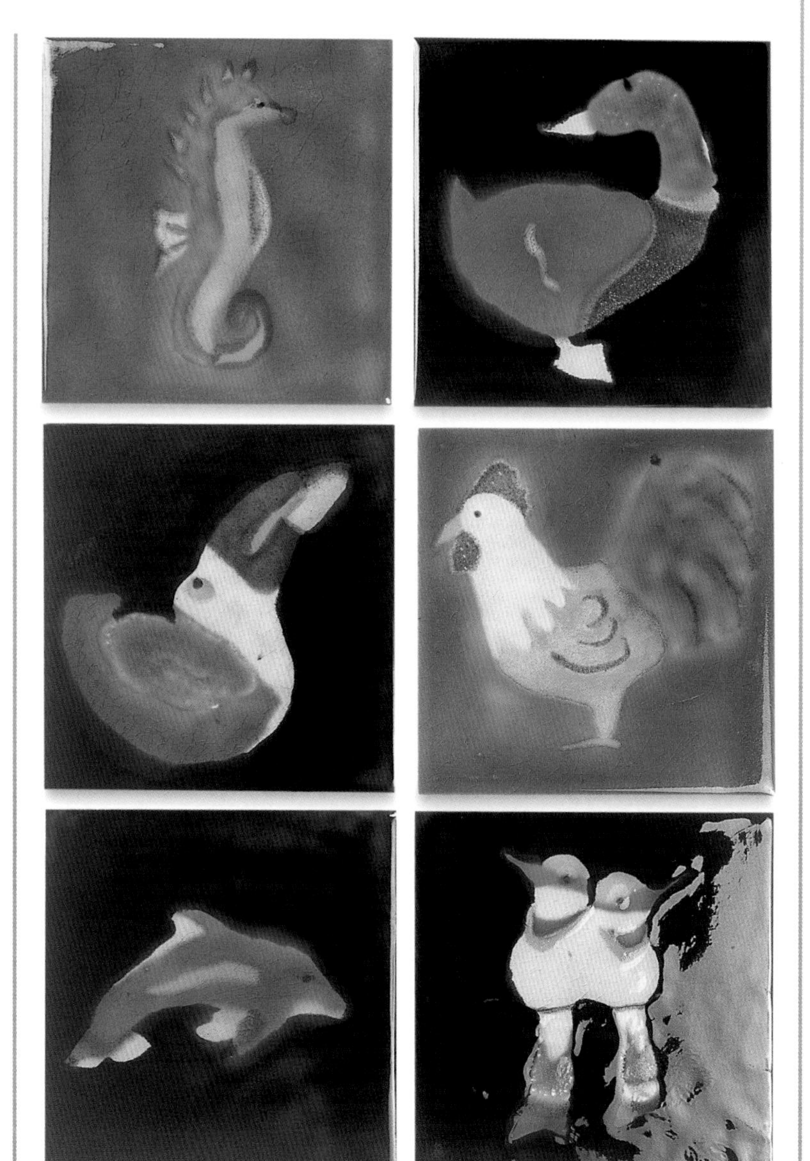

NOREEN JAAFAR

Birds and marine animals
The artist of this collection finds that aquatic themes are among the most popular for tiles. Molten cobalt and copper are used to give a glossy, stained glass effect. Because of their intense color, these tiles look good in a bathroom and can cover a large area of wall.

◀ CORRES MEXICAN TILES

Blue vase
These delicately drawn Mexican tiles, with their finely detailed blue birds perching among mimosas and heavy blue daisy heads in a blue and white vase, are complemented by the painstakingly detailed border tiles, creating a charming mural. The butterfly adds a lively touch. The base tiles are handmade earthenware, handpainted and overglazed in Mexico, using traditional designs.

Working with Tiles

The advice that follows will help you to plan a tiling scheme and carry out all the practical work involved, from design and preparation to fixing tiles on walls and surfaces.

Before starting any tiling job, it is essential to have the right tools and materials. Always buy the best tools you can afford, keep them clean, and put them away carefully. Expensive specialized tools, such as heavy floor-tile cutters and cutting jigs, can often be rented.

Tools and Materials

If you need to undertake major renovation before you start tiling, relaying a floor, removing old tiles, or replastering walls, you will need additional tools for these jobs.

Tools for planning

1 Retractable metal tape measure, preferably 25 feet (7.5m) long; **2** 3-foot (1m) heavy metal ruler; **3** a drawing compass; **4** pencil; **5** pen or ballpoint pen; **6** colored felt-tip pens or crayons; **7** eraser; **8** sketch pad; **9** graph paper; **10** tracing paper; **11** pocket calculator (optional).

Tools for tiling

12 Plumb line; **13** carpenter's level; **14** carpenter's square or T-square; **15** pencil; **16** felt-tip pen; **17** wooden strip or lath; **18** wooden marking stick, small hammer; **19** nails; **20** cardboard (for templates); **21** metal or plastic template profile marker;**22** tile cutter or cutting jig; **23** tile-edge sander; **24** pliers or special tile nibbling pincers; **25** tile saw (optional); **26** vise or adjustable workbench; **27** padding; **28** notched adhesive spreader; **29** rubber-blade grout spreader; **30** rounded stick or grout joint finisher (for smoothing grout joints) or $\frac{1}{4}$-inch (6mm) wooden dowel for floor tiles; **31** bucket; **32** sponge; **33** cloth.

Materials

34 Tiles; **35** border tiles; **36** quadrant tiles for corners, glazed tiles for edges and nosing (if required); **37** sealing strip (for positioning between wall tiles and countertop, bathtub, or sink); **38** spacers (or matchsticks); **39** adhesive (waterproof adhesive is essential for bathrooms and kitchens); **40** grout (special epoxy grout is needed for kitchen counters); **41** color for grout (if required).

TIPS

- Always buy or make enough materials to finish the job in one operation; nothing is worse than running out of materials halfway through.
- If you are using plain tiles in one color, make sure all the tiles come from one firing; otherwise, there may be color differences that will show up when the tiles are in position.

Tiles have a number of design possibilities. Their simple, geometric shapes lend themselves easily to patterns and designs, and these designs can be used to create interest in the home or change the feel of a room. Consider the options before making a decision.

Design Considerations

ABOVE Set of tiles designed as a picture for a child's bathroom.

One of the simplest ways of breaking up a run of plain tiles on a wall is to drop in motifs or picture tiles. This can be very effective, but these tiles need to be placed carefully.

Another exciting use of tiles, particularly in children's bathrooms, is to decorate a wall with a set of tiles designed as a picture, for example, a scene from a nursery rhyme or children's story.

Color and pattern

When choosing tiles, remember that color and pattern help to create decorating illusions and can be used to set the style and alter the shape and proportion of any room: warm-colored, textured tiles will make a large bare space look smaller and more intimate; highly glazed pale tiles that reflect the light can make a small space appear bigger.

Tiles laid on the floor in two or three different colors in a regularly repeating sequence can suggest a giant board game; checkerboard patterns, traditionally black and white, have a timeless appeal; and you can use darker or brighter tiles to make a border effect on the floor or on a wall.

Glazed tiles with a high gloss finish become too slippery when they are wet to use as an entire floor covering. They can, however, be safely inset amongst large unglazed terracotta tiles like small bright gems set in a dull stone matrix. The unglazed tiles can be made octagonal, or their corners can be cut off to accommodate the

ABOVE Three colors used in a regularly repeating sequence.

BELOW A patchwork panel above a sink.

ABOVE Vary the texture of a wall by turning square tiles upright like diamonds.

glazed insets. The glazed tiles should be no larger than 4 x 4 inches (100 x 100mm) and they should be decorated with a symmetrical design that can be read from any direction.

Patchwork effects, made by adding the occasional patterned or pictorial tile to a plain wall, can look very good in a small room, or on a wall above a sink, and you can vary the texture of a tiled wall by hanging a square tile on point like a diamond, filling the spaces with triangles made by cutting two tiles in half diagonally. Colored grouting can enhance any patchwork effect and can be used to emphasize any tiles that are hung this way.

It is a good idea to test any design ideas on site, and it is worth making sample tiles to make sure the pattern you plan works. Nothing is worse than creating a design that you decide you don't like, or others in the family don't approve of, and the same goes for colored grouting. Test a sample first before using it on a wall.

Borders

One of the easiest things to do with tiles is to create a border. This can be used either to frame a feature, or to suggest a natural division in a room. Heavier relief-type border tiles look particularly effective finishing off a tiled dado. Always check that the border tile you make is the same width as the main tile.

Border designs

The simplest way to make a border is to run a row of plain tiles along the bottom of a surface in a contrasting color. These colored tiles can be solid or alternated with the tiles used on the walls. Another idea is to incorporate alternate patterned tiles, and it is also possible to buy or create strongly patterned tiles in bold primary colors, such as a child's alphabet, for a child's room or bathroom.

Border treatments of this type do not have to be confined to walls. Floor tiles can also have borders to define areas and outline features.

Plain floor tiles can have patterned borders, and this was often the case in Victorian households. A common treatment was to alternate red and black tiles and surround them with a patterned border of encaustic tiles in matching colors.

If the whole of the downstairs area is to be tiled in the same color – a design trick that creates the illusion of space in a small house – border tiles can be used to separate different areas, or at the entrance to rooms. Tiles also make a wonderful threshold at door openings and provide a good visual link between two disparate floorings.

ABOVE For a child's room, use strongly patterned tiles in bright primary colors.

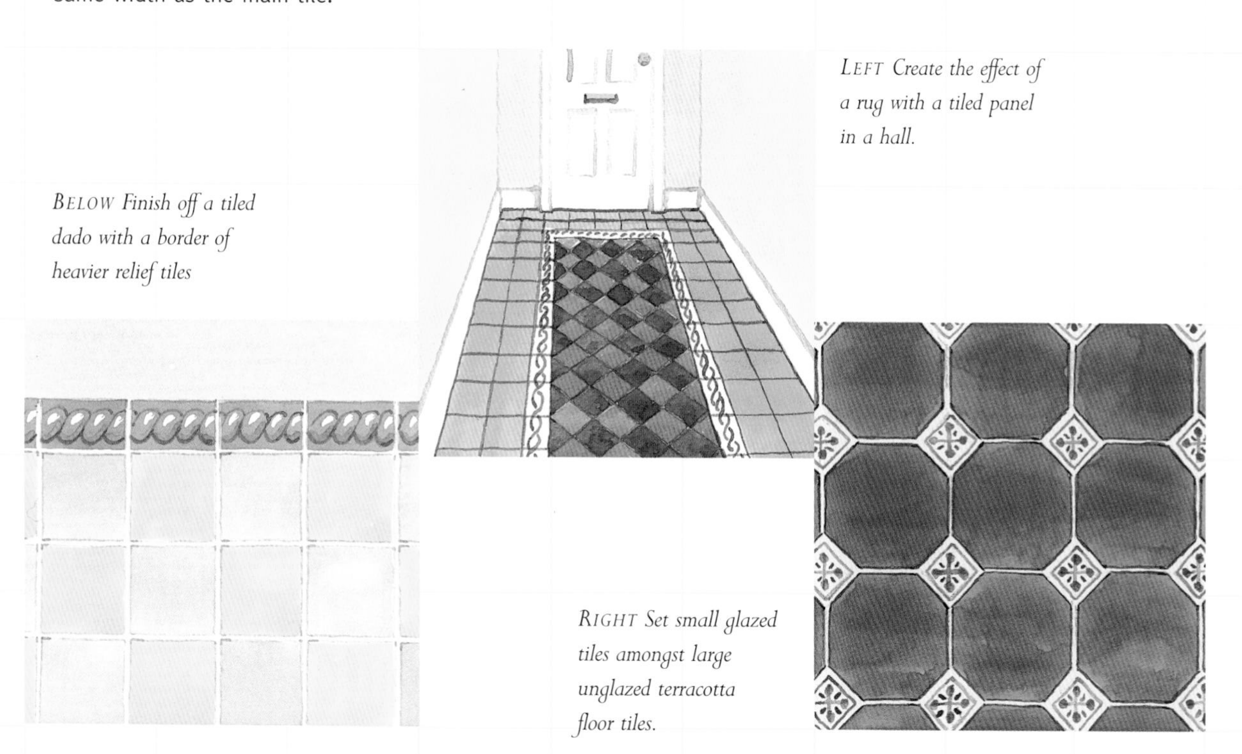

LEFT Create the effect of a rug with a tiled panel in a hall.

BELOW Finish off a tiled dado with a border of heavier relief tiles

RIGHT Set small glazed tiles amongst large unglazed terracotta floor tiles.

Good planning and close attention to the practical aspects of your tile design are the keys to success. By drawing an accurate scale plan you can identify any potential problems and make the necessary adjustments before you begin tiling.

Planning the Design

The first thing to do is to choose the right size of tile for the surface you plan to cover. Small tiles are best in small rooms or on small areas, such as kitchen counters, and around sinks in bathrooms. Here large tiles would be awkward and have to be cut to fit. On larger areas, such as walls around showers or on floors, you need to use larger tiles; otherwise, the laying process becomes too complicated, and it can be difficult to plan any pattern successfully. Use border tiles if you want to outline a feature or trim a dado or a half-tiled wall area. This can make a space look longer, wider, or smaller.

Once you have decided on the size and type of the tile you are going to use, there is only one successful way to plan how to tile an area and to judge any design and that is to put the design accurately on paper. A good plan allows you to make any adjustments necessary, and you can use it to calculate the number of tiles you need accurately.

First, make a rough plan of the room you are going to tile, including any features such as bathtubs.

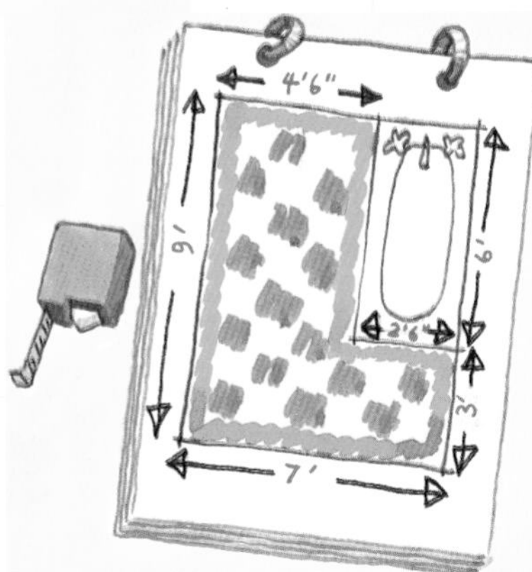

Work out approximate measurements and mark these on your sketch.

Making the plan

Make a rough plan of the floor and walls you are going to tile on a sketch pad. Follow the shape of the room. Then make a rough sketch of your design. When you are satisfied, measure the floor and walls accurately and prepare a basic plan. Do this even if you have architect's or builder's plans. Start with the floor, draw the shape of the room in the chosen scale on graph paper, making one square equal to one unit of measurement. If you are working in standard measure, make one foot of floor space equal one inch or half an inch on the paper. If you are using metric measures, a scale of 1:20, 1:25, or 1:50 is usual.

Draw all the recesses and projections to scale using a compass for the curves and a square for the right angles. Mark the exact position of all the doors and windows and draw the swing of the doors and windows (the way they open). Check that the corners and walls are vertical with a plumb line.

Then make plans of all the walls that are to be tiled using the same scale and draw accurately the position of the windows and doors and any other items on the wall, such as built-in furniture or bathroom or kitchen appliances, including all the electrical outlets. It is simplest to work on graph paper, but if you want to check the design on a larger scale, transfer your measurements on to a large sheet.

Working out the design

Make several copies of your accurate scale plan. Draw the area that you plan to tile and then mark it accurately into squares, making each square the correct size for the tiles that you are going to use. Mark in the position of any colored or decorated tiles with colored pens or pencils. Look at the finished design and compare it directly with the wall in the room. Discuss the design with all the other people who will use the room. Remember that any pattern or motif should be centered on any architectural feature, such as a mantelpiece, in the same way as you would with wallpaper.

When you have worked out the design satisfactorily, then transfer it carefully to a tracing paper overlay. This can be held in place over the floor or wall plan and will help you to position the tiles to avoid difficult edging or any awkwardly placed motif.

If you have planned your design accurately, you can count the exact number of tiles that you are going to need from the design. Check the measurement of the size of the tiles one final time just to make sure that you haven't made a mistake, and allow some extra tiles for any that may be damaged when they are cut to fit.

TIPS

- If you are planning a pattern, invest in one or two sample tiles to check the colors on site.
- As a general rule, position tiles so they are centered on the wall or on the main feature, such as a sink or fireplace.
- Check your design by laying out the tiles in sequence on a table or on the floor. Often a design appears much brighter and bolder with the tiles themselves, and it is better (and cheaper) to change it before the tiles are set in place.
- Check to see that any pattern is not broken up awkwardly when you reach a corner or projection and adjust the positioning if necessary.
- All tiles for places where there is water, particularly bathrooms and kitchens around the sink, must be waterproof.
- Textured floor tiles are good for areas of heavy use, but they can be difficult to clean and are not really necessary in an ordinary house.
- If you are tiling a porch, patio, or sunroom, make sure you use frostproof tiles suitable for outdoor use, and the same applies to paths and areas around swimming pools.
- If you want to use tiles to surround ovens, around fireplaces, or at the back of stoves, where they will have to withstand excessive heat, make sure the tiles are strong enough and of sturdy enough manufacture. Many special fireplace tiles are produced that are intended as a fill-in between the grate and the fire surround.
- If you want a tiled hearth, the tiles need to be heatproof. Floor or quarry tiles are often used. Choose tiles that harmonize with any tiles used in the fireplace area.

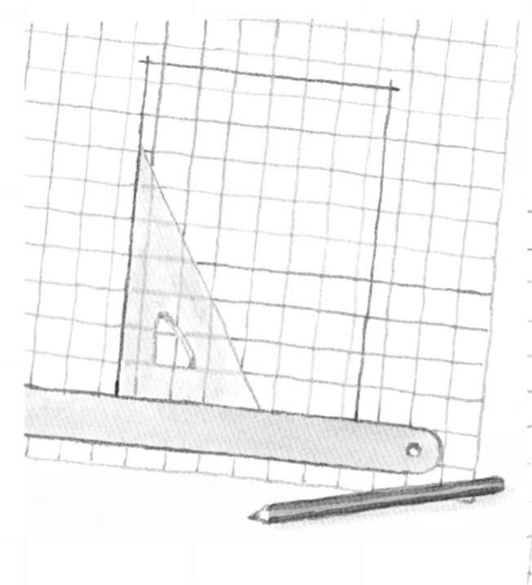

ABOVE Draw an accurate scale plan on graph paper, using a square and ruler.

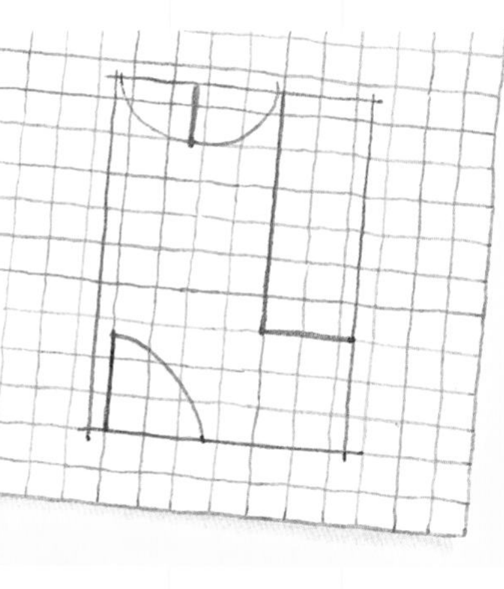

ABOVE Draw in all recesses and projections, and mark which way doors and windows open.

BELOW Mark the area you plan to tile into squares, coloring in your design with pens or pencils.

If you know the basic techniques for working with tiles, you can achieve professional-looking results. The guidelines given below on preparation and procedures will help you to start with confidence and avoid many of the problems that commanly arise.

Practical Tiling

Plan carefully before you start tiling. In wet areas, such as bathrooms, check that no moisture will be trapped in the wall behind your tiles.

Before you contemplate any tiling at all, it is essential to check that the structure of the building is sound. If you are going to tile a wall, it must be firm and dry, for if the plaster or wallboard is in bad condition, the tiles may just fall off. Don't consider any tiling, either of floors or walls, if any condensation is present, particularly in wet areas such as bathrooms or utility rooms. Tiles form an impermeable barrier, and if there is moisture present in the wall behind the tiles, this could cause the wood of the building to rot.

Tiles are also very heavy. If you are planning to lay tiles on a floor in an upstairs room, or where there is an existing suspended wooden floor, you must check that joists and floorboards are strong enough to take the weight. Joists tend to get weaker as they go up the house. If necessary the joists will have to be strengthened. Take specialist advice if you are in doubt.

Advance planning

It is also very important to think ahead about any future structural work you may want to do on the house. New lighting and electrical works may require cables to be embedded in walls, and any new plumbing may require new pipe runs. If you are laying a tiled floor, try not to embed cables or pipes in the floor, for if something goes wrong you will then have to excavate the whole floor to find the fault. It is better to run the pipes and cables around the edge of the room and cover them with baseboards, or box them in.

In the same way, try not to embed cables in the wall behind the tiles and try to avoid positioning switches in a tiled area. If this is impossible, consider installing an inspection hatch which can be tiled to match the wall. Similarly you can install a panel over the bathtub, tiled to match the rest of the bathroom, that can be removed without breaking the tiles if any plumbing problems arise.

Preparing walls for tiling

Before starting, make sure all surfaces are clean and in good structural repair. Remove any hooks or nails, pull out any old hardware, sand down bumps, and fill in any holes or depressions. Sand down the wall until it is level and seal newly filled areas with tile adhesive primer.

Newly plastered walls Leave new plaster for at least one month to dry out and seal with tile adhesive primer.

Old plaster Examine the wall carefully and fill any cracks. Let the putty dry. Sand the wall to remove any bumps. Apply tile adhesive primer to any new areas of plaster as above.

Wallboard Wallboard for tiling must be firm and secured on a proper framework which does not flex. It is not suitable for heavy tiles. If the wallboard is new, treat it as plaster (see above). If it has been painted, treat the surface as paint (see below).

Painted walls Sand all painted surfaces, especially anywhere the paint is flaky or has been damp and wash or wipe down the surface. Leave it to dry. Gloss paint will need to be sanded until it is matte and then washed. Score all painted surfaces with the corner of a paint scraper in a crisscross pattern to allow the tile adhesive to grip.

Wallpapered walls Do not tile directly onto any wall which has been covered with wallpaper. Remove the old paper and clean and prepare walls in the normal way.

Paneled walls These are usually not suitable for tiling. Special panels can be made from blockboard or water-and-boil-proof (WPB) plywood that can then be tiled for boxing in pipes in bathrooms, or for access to the wall behind the tiles. Bare wood must be primed with an oil-based primer. Chipboard and insulation board, which swell when wet, and masonite, which is too flimsy, are not suitable for tiling over.

Existing tiles It is possible to tile on top of existing tiles, but you must make sure that they are firmly stuck to the wall as any heavy tiles could pull the old ones away. It is best to use self-spacing Universal tiles with beveled, glazed edges, as these are easier to hang on previously tiled walls. Note that tile adhesive will take longer to dry than on a plaster wall, and the tiles need to be left for 3–4 days before grouting.

Preparing floors for tiling

If you can afford it, it is best to employ a professional flooring contractor to lay floor tiles. As with all jobs, get estimates from two or three suppliers to compare prices and ask for recommendations from satisfied customers.

If you are planning to lay the flooring tiles yourself, make sure that all floors are clean, smooth, level, and free from moisture. Check with a carpenter's level that the floor is even. If it is uneven, you may need to apply a floor-leveling compound first.

Cutting tiles

To cut wall tiles accurately, first measure the area that you want to cut and mark it on the glazed side of the tile. Never mark the back of a tile with a felt-tip pen or marker as the clay absorbs the ink, which then seeps through to the front of the tile and usually appears as a stain after hanging. Score along the tile using a proper tile cutter and metal ruler. Place a matchstick under the tile along the scored line and then press firmly down on both sides of the tile. This will break the tile cleanly in two. Special tile-cutting jigs are available with built-in marking gauges if you want to invest in one. This makes the job easier.

Cutting corners and angles

You will need to cut corners out of a number of tiles to fit tidily around electrical switches or other fixtures.

Mark and score the tile as before, taking care not to score into the surface of the tile that you are retaining. Then "nibble" away the unwanted area with a pair of pincers or pliers. Alternatively use a special tile-cutting blade in a hacksaw and smooth any rough edges with a carborundum stone or tile file.

Cutting floor tiles

These are much harder to cut than wall tiles since they are that much thicker, and an ordinary wall tile cutter is not strong enough to cope. You may be well advised to get specialists to cut and lay floor tiles for you.

SAFETY NOTE
Glazed tiles are vitrified, and the edges can be very sharp when they are cut. Small splinters of glass can also break off. Always wear gloves and goggles when cutting or drilling tiles.

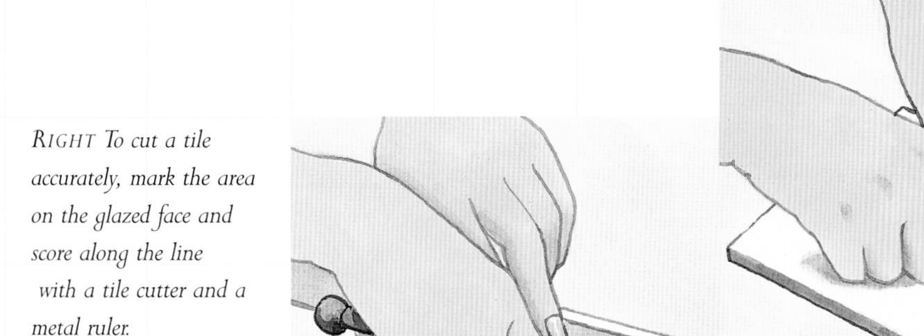

RIGHT To cut a tile accurately, mark the area on the glazed face and score along the line with a tile cutter and a metal ruler.

ABOVE Place a matchstick under the scored line and then press down firmly on both sides of the tile to make a clean break.

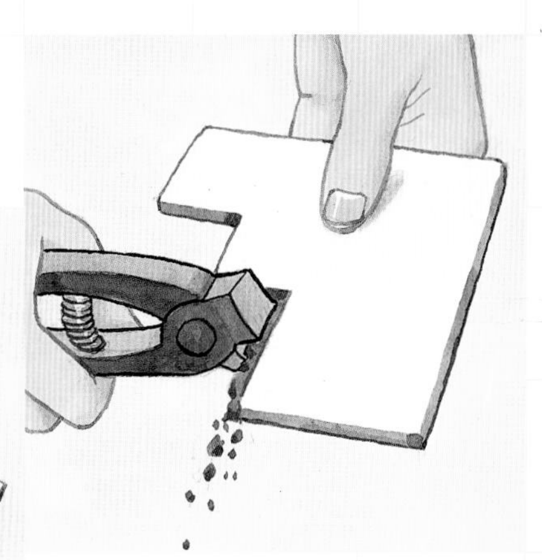

ABOVE Cut out corners by "nibbling" away unwanted areas with a pair of pincers or pliers.

Tiling walls and floors

Find the exact midpoint of the wall, floor, or area to be tiled. Use a plumb line and level on walls to make sure the vertical and horizontal measures are accurate.

If you have not planned your design in detail, lay out a row of tiles as a dry run, placing one tile on each side of the central line. If this arrangement proves awkward and results in difficult cuts at each end of the row of tiles, center one tile on the midpoint and work out from this.

Mark off the width of the tiles you are using on a length of wood which you can use as a gauge or marking stick. Use the stick to test various starting points. Make two measuring sticks if you are using rectangular tiles.

It is essential to establish a level base since the bottom row of tiles on a wall must be even. This means drawing a true horizontal line since the baseboard or floor may not be horizontal. Use a level and strip of wood to establish the true horizontal, and draw a line or tack a strip of wood all along the wall or around the room if you are tiling more than one wall. Draw this line approximately one tile's depth above the baseboard, but make sure the line is never more than one tile's depth above the baseboard. Redraw the line or reposition the strip if necessary.

ABOVE From the midpoint of the area to be tiled, check your vertical and horizontal measurements with a plumb line.

ABOVE Mark off the width of your tiles on a length of wood and use this as a gauge or measuring stick.

Check that the lines meet exactly at the corners if you are tiling a whole room, and try to ensure that tiles line up with window frames and are level with any prominent features.

Laying the tiles

When you have finally decided on your starting point, mark it carefully and start laying the tiles. Always use a recommended tile adhesive and a good notched spreader. A waterproof adhesive is essential for kitchens, showers, and bathrooms.

Do not try to cover too large an area at once. Spread the adhesive and lay the tiles in blocks of 1 square yard/meter at a time. It is best to start in the midpoint, but you can start in a corner if you are sure that the corner is exactly square and there are no tile cuts.

Spread the adhesive about 1/16 inch (2–3mm) thick and comb it through with the notches on the spreader.

Once you have spread the adhesive, lay the first tile. Check that it is absolutely square with a plumb line and level. If you are using Universal tiles, you will not need to use matchsticks or tile spacers, but these will be necessary if the tiles do not have glazed or beveled edges. Push the spacers into the adhesive between the tiles. Position the tiles in rows and check periodically that they are square, level, and flush.

Continue spreading and laying until you have filled the area that is to be covered with uncut tiles. Leave for 12–24 hours for the adhesive to set. Remove the batten, if you have one in place, and pull out the spacers. Measure, cut, and lay the tiles that are needed at the edge of the tiling (if necessary), and if you are using border tiles, lay them in position. Once the adhesive has set, you can clean and grout the tiles.

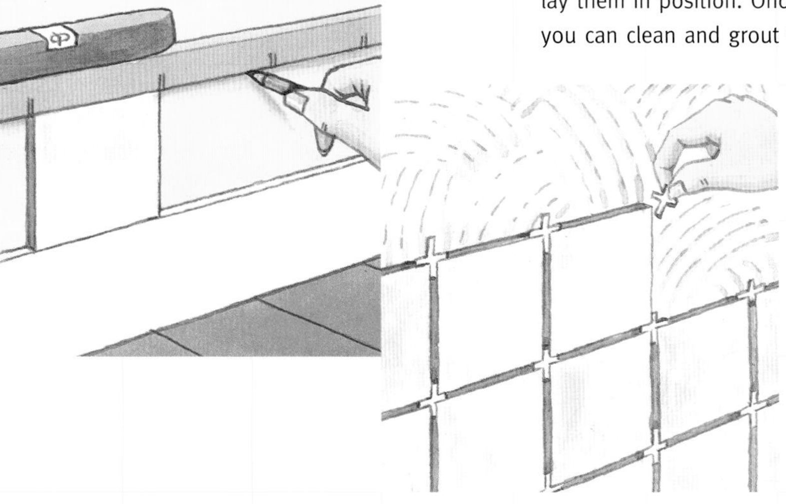

RIGHT Establish a true horizontal line at the base of the wall by using a strip of wood and a level.

If your tiles do not have glazed or beveled edges, you will have to push tile spacers into the adhesive between them.

Tiling surfaces

The principle of tiling counters is exactly the same as tiling a wall. Tiles are heavy, so first of all check that any furniture or kitchen cabinet is strong enough to take the weight. Strengthen the joints and frames if necessary. For kitchen counters and kitchen and dining-room tables, you need to use tiles that are at least ¼ inch (6mm) thick to withstand the heat of any heavy pans or dishes. They should also not be too highly glazed since this produces a slippery surface. Surfaces need to be finished off neatly using special corner and edging tiles; the top can be finished and edged with hardwood, which can either be stained or painted to match the decoration in the room.

Sizing up

The size of the tile that you use is generally a matter of taste, but it often helps to measure the surface first and then work out the tiling patterns with various sizes of tile. This may help to avoid cutting too many tiles, and it is always better to try to cover the surface with as many uncut tiles as possible. Plan the layout, and position any tiles that have to be cut at the back of the work surface. If you are tiling an L-shape, make sure that a whole tile meets its partner at the inner bend of the "L" and that seams are continuous along both surfaces.

If you are tiling a square or rectangular table and you are going to have to cut tiles, work from the center out so the cut tiles fall at the edge and the table looks symmetrical.

Laying the tiles

When you are tiling the main part of the top, spread the adhesive on the prepared surface with a notched spreader. Lay 4–6 tiles at a time. Spread the adhesive ⅛ inch (3mm) thick. Push the tiles firmly into position and check that they are flush and level. Insert spacers between each tile. When the main tiles are in position, cut any tiles to size that are needed to fill in the edges and round the sides. If necessary, make a template for any awkward shapes.

Leave the adhesive to harden and then apply the grout. Epoxy grout (see **Tips**, page 138) is sticky and different from conventional grouting. Follow the manufacturer's instructions carefully.

Decorative edging tiles look particularly effective when used to finish a counter. Make sure they will withstand the heat of the pans and dishes.

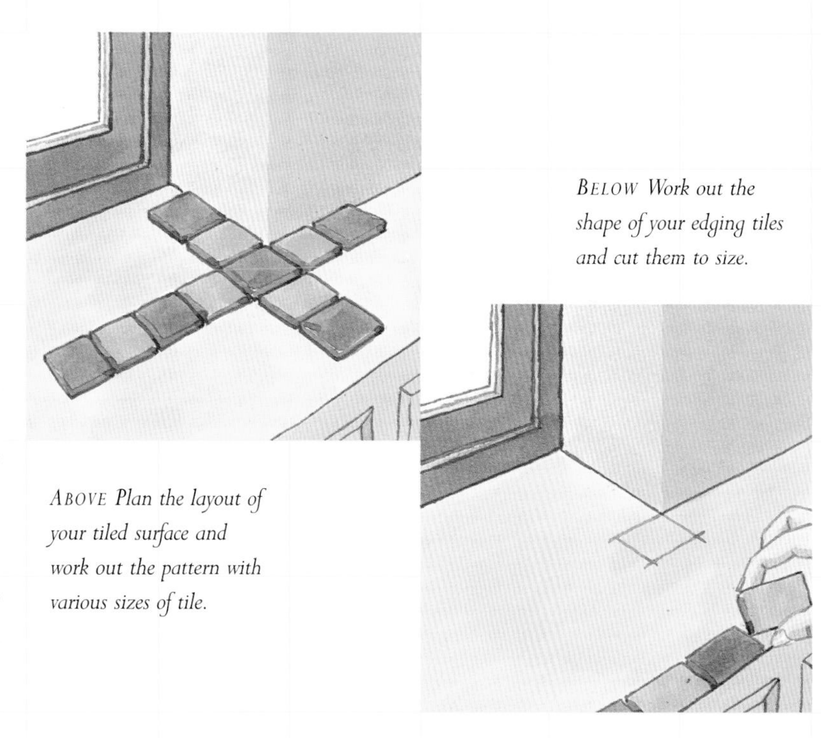

ABOVE Plan the layout of your tiled surface and work out the pattern with various sizes of tile.

BELOW Work out the shape of your edging tiles and cut them to size.

Grouting

Once the tiles have been stuck in position for about 12–48 hours and the tile adhesive is set, the gaps between the tiles must be grouted to prevent water and other liquid and dirt from seeping in, which in time would make the tile drop off.

Most grout is neutral-colored, but colored grout can be very effective if you want to emphasize the lines made by the tiles, or you want the grout to blend in with the color of the tiles. Colored grouting is available in a limited choice of colors, but you can color standard white grouting with poster paint or acrylics.

Grouting is a messy job, so the first thing to do is to cover all the surface surrounding the tiles before you start. This applies particularly to walls, since grout tends to drip down onto floors or into bathtubs.

Grout comes either in powdered form or ready-mixed. Mix the grout if you are using powdered grouting and in any case give the grout a good stir. Make sure you have enough, but it is often better to mix powdered grout in batches as it hardens quickly, and if you have a large job to do, some of it may be wasted.

Above Spread the grout over the surface, working quickly and pressing it well into the gaps between the tiles.

Spread the grout over the surface with a special grout spreader, a rubber squeegee blade, or a damp sponge. Work quickly, pressing the grout well into the gaps between the tiles. Wipe away all the excess with a damp sponge as you go along and don't let the grout harden on the front of the tile as it will be difficult to remove. Wash out the sponge frequently; otherwise, you will leave smears on the tiles.

As the grout begins to harden, run a rounded stick between the joints to make a smooth, even surface. Check that there are no gaps, and wipe the front of the tiles carefully if necessary.

Don't grout along the space where the tiles butt up to an adjoining surface, such as a bath-tub and scrape away grout that gets into this gap. It will need to be sealed with a silicone or acrylic sealant after the grout has fully dried.

GROUTING TIPS

- Always make sure you have the right grouting for the job. A special epoxy grout for countertops is essential as it is impervious to cooking spills and acids. With handmade tiles, use a grout formulated for wider joints.
- You can regrout an existing tiled wall using a colored grout. This will change the look of the tiles completely and give the room a new lease on life.
- If you have persistent grout stains on a tiled floor after grouting, these can usually be removed by rubbing the floor with sawdust. Use a damp cloth to rub the sawdust in. It can then easily be swept up.
- Replace the lid on the grouting when you are not using it. This prevents it from hardening and being wasted.

Wipe away all excess grout with a damp sponge as you go along, otherwise as it dries it will be difficult to remove from tile surfaces.

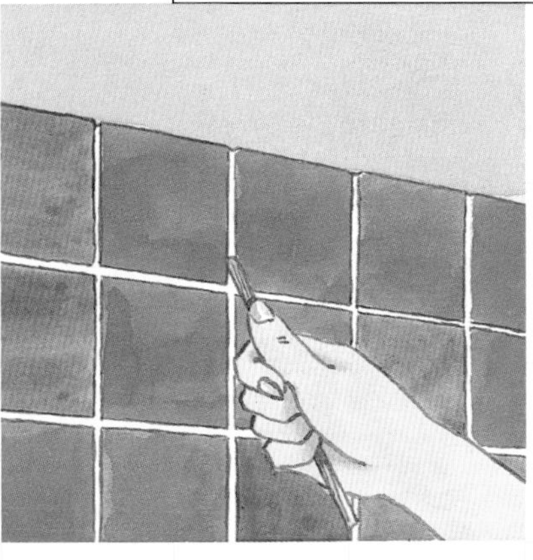

When the grout begins to harden, run a rounded stick between the joints to make an even surface.

Additional Equipment

The instructions given below are for the construction of simple items of equipment for use with two of the projects in this book.

Equipment & Materials

- 5 feet (1.5m) of 2 × ⅔ in (50 × 16mm) finished lumber
- Handsaw
- Wood plane
- Wood glue
- 3 × 3 feet (1 × 1m) of ½-inch (12mm) plywood base board
- Electric drill with ¼-inch (6mm) bit
- Hacksaw
- 12 inches (300mm) of ¼-inch (6mm) diameter brass rod
- Absorbent board or canvas cloth
- Soft brush
- Talc

OPEN-FACE MOLDED TILES

(see pages 75–7)

Making the frame

1 Cut the lumber into four 12½-inch (318mm) lengths with the handsaw and plane a ⅛-inch (2mm) chamfer on one edge of each. Using wood glue, carefully miter the four pieces of wood together to make a square frame, making sure the narrower face of each is uppermost. The opening framed by the wood should be 8½ × 8½ inches (218 × 218mm) at the bottom and 8¾ × 8¾ inches (222 × 222mm) at the top.

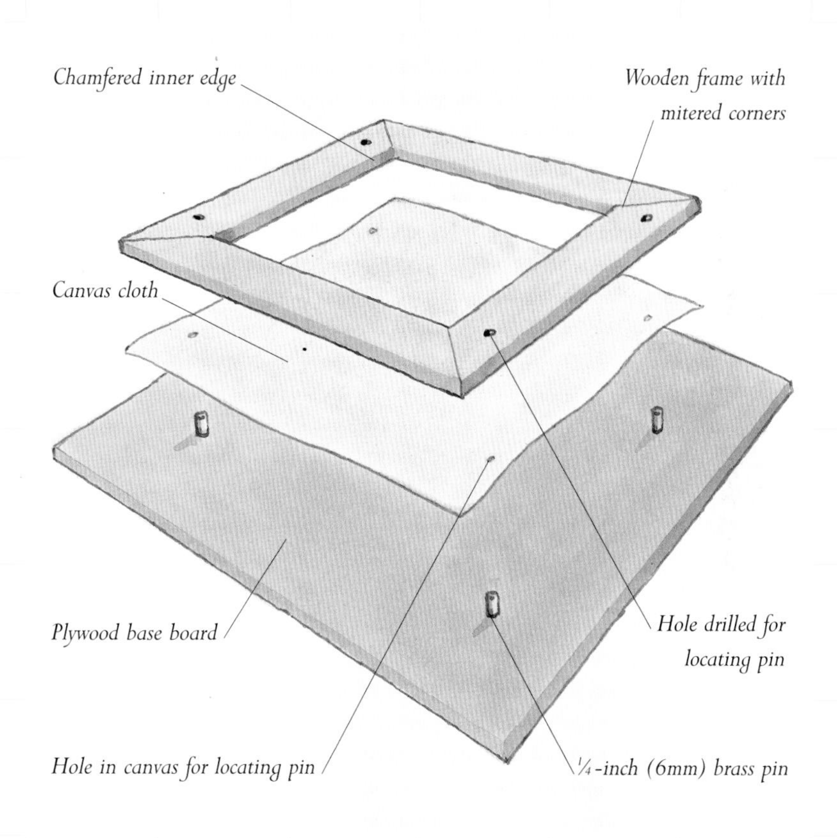

2 Position the frame on the plywood base board, and drill a ¼-inch (6mm) hole cleanly through each corner and the underlying board. Make sure the drill is kept as vertical as possible. With the hacksaw, cut four short pins from the brass rod and drive them into the holes in the board. (These will secure the frame in position while clay is thrown into it.) Place a piece of absorbent board or canvas between the pins and locate the frame over them. Make sure the pins do not protrude above the frame's upper surface. Using a soft brush, lightly dust the interior of the frame with talc.

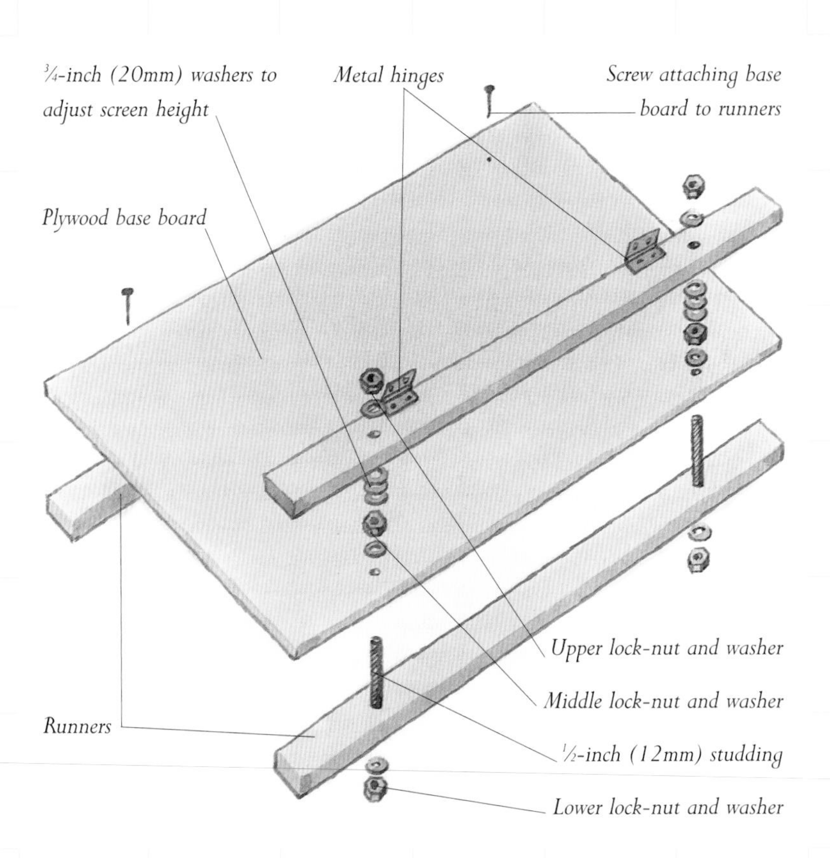

Equipment & Materials

Handsaw
9 feet (2.7m) of 1 × 2-inch (25 × 50mm) lumber
Small screws
24 × 36 inches (600 × 900mm) sheet of ½-inch (12mm) plywood
Electric drill with ½-inch (12mm) and ¼-inch (5mm) bits
2– 4-inch (100mm) lengths of ½-inch (12mm) studding
8 lock-nuts
8– ¾-inch (20mm) washers
2 metal hinges
2– ¼-inch (5mm) threaded bolts with 8 nuts and washers
Small piece of wood for screen prop
Sheet of thick glass
Thin wooden beading
Brads

SCREEN PRINTED DESIGNS
(see pages 114–19)
Making a printing table

1 Cut two 36-inch (900mm) lengths from the wood and screw them to the underside of the plywood base board on the long edges. Through one of these runners and the base board, drill two ½ inch (12mm) holes approximately 6 inches (150mm) in from each end. (These are to take the two pieces of studding.) Each hole should be counter-sunk to house a ¾-inch (20mm) washer and lock-nut. The studding will be locked into place with a second nut and washer screwed down onto the base board from the top.

2 Cut another 36-inch (900mm) length of wood and drill two ½-inch (12mm) holes to correspond with the first pair so that this piece of wood can ride up and down on the studding. Its height off the base board must be variable to adjust the screen distance from the printing surface, and it will be held at the correct height with lock-nuts on each side. Screw two metal hinges to this wood to attach whichever screen you wish to use when you are ready to print.

3 On the opposite side of the screen from the hinges, drill two holes through the frame to take ¼-inch (5mm) threaded bolts. These are used with the addition of small nuts and washers to adjust the height of the frame off the base board to correspond with the hinged side.

4 With a screw and washer, fasten a small wooden prop to one side of the screen to support it when it is swung away from the printing surface. Finally, position a thick sheet of glass as a base to print on directly below the stencil. When you set up each print run, the glass can be held in position with a thin piece of wooden beading tacked to the base board with small brads.

Glaze Recipes

Glazes are generally readily available from commercial suppliers. The following recipes are suitable for use with certain projects in this book if you wish to mix your own glazes or are unable to obtain the medium you specifically require. Cobalt lining pigment cannot be purchased ready mixed.

When handling any of the materials listed below, take the appropriate safety precautions as advised by the suppliers or manufacturers.

All proportions given in the recipes are by weight and can be converted easily to actual measurements by assuming each percentage equals an ounce or gram.

MAJOLICA PIGMENTS
Ingredients:

Black
- 2 parts black iron oxide
- 1 part nickel oxide
- 1 part china clay

Green
- 4 parts copper carbonate
- l part nickel oxide

Yellow
- 2 parts lead antimonate
- 1 part lead sesqui-silicate

Orange
- 1 part orange glaze stain
- 1 part lead sesqui-silicate

Raspberry red
- 1 part chrome tin pink
- 1 part bismuth oxide

ROLLED TILES (pages 70–2)
Turquoise earthenware glaze (Orton cone 03)
Ingredients:
- 40% ferro frit 3110
- 30% standard borax frit
- 10% zinc oxide
- 10% flint
- 5% ball clay
- 3% zirconium silicate
- 2% black copper oxide

To help keep this glaze suspended without increasing the clay content, white craft glue can be added to the glaze slop. The more clay you add, the more the color changes from turquoise to green.

Clear earthenware glaze (Orton cone 03)
Ingredients:
- 70% lead sesqui-silicate
- 10% ball clay
- 10% flint
- 5% china clay
- 5% whiting

To create the colored glazes for splashing over this, add 3% copper carbonate for a green or 4% black iron oxide for a honey brown.

DELFT DESIGN (pages 86–8)
Tin glaze (Orton cone 03)
- 54% standard borax frit
- 28% lead bisilicate
- 9% tin oxide
- 9% red clay

Tin glazes can only be applied to biscuit-fired ware as the clay content of the glaze must be kept low.

Tin oxide, which creates the characteristic whiteness and opacity of a tin glaze, is an expensive material. It can be replaced with zinc oxide to achieve similar effects, although zinc glazes are slightly harsher in appearance and have a duller color response to some of the majolica pigments.

Cobalt lining pigment
Ingredients:
- 4 parts cobalt oxide
- 2 parts china clay
- 1 part black iron oxide

These should be ground together with water using a pestle and mortar. To help the pigments flow easily when you are painting, you can dip the brush in water mixed with gum arabic.

Index

All techniques and projects are illustrated. Page numbers in *italics* refer to illustrations in the Galleries and Introduction.

Acknowledgments

Quarto would like to thank the following suppliers for kindly lending us items for use in photography:

Cerametech Ltd, 16-17 Frontier Works, 33 Queen Street, London N17

David Mellor, 4 Sloane Square, London SW1W

Tower Ceramics, 91 Parkway, Camden, London NW1

We would also like to thank:

Claire Godsell, who designed the motif used in the project on pages 81-3; Dorothy Frame (indexer); Barty Phillips (additional information).